STATISTICS MADE LEARNABLE

RICHIE HERINK

Fideli Publishing, Inc.
119 W. Morgan St.
Martinsville, IN
46151

3rd Edition

ISBN 978-1-60414-607-3

\

CONTENTS

CHAPTER 1
Averages....... 1
CHAPTER 2
Standard Deviation....... 5
CHAPTER 3
Frequency....... 10
CHAPTER 4
Assumed Average 15
CHAPTER 5
Grouped Data 17
CHAPTER 6
Frequency Distribution 26
CHAPTER 7
Permutations and Combinations 30
CHAPTER 8
Probability....... 36
CHAPTER 9
Binomial Distribution....... 40
CHAPTER 10
Normal Distribution 44
CHAPTER 11
Poisson Distribution 52
CHAPTER 12
Sampling Theory 56
CHAPTER 13
Difference Between Two Means....... 58
CHAPTER 14
Analysis of Variance....... 63
CHAPTER 15
Coefficient of Correlation 70
CHAPTER 16
Chi Square Test....... 75
CHAPTER 17
Linear Regression 79
CHAPTER 18
Index Numbers 84

Index 89

A NOTE FROM THE AUTHOR

Thanks to the numerous college professors throughout the country and overseas who requested that this type of learning aid for students be published as both a paperback book and an e-book to augment the textbook and reference materials that they are using for their own online and off-line Statistics courses.

The numerous helpful suggestions received from these teachers have assisted in producing a well organized set of notes that will give the student a fuller understanding of the material being presented.

Only simple math is used and straightforward step-by-step examples significantly reduce the difficulty of learning the more complex Statistical concepts.

The English used is simple and straightforward thereby, making this learning aid especially useful for facilitating teaching the exact same Statistics course online and worldwide.

In addition, the e-book version of this publication can be used to supplement appropriate MOOCs — massive online open courses.

Richie Herink, PhD

Ridgewood, New Jersey

1 AVERAGES

HOW TO CALCULATE THE MEAN, MEDIAN, MODE AND RANGE FROM UNGROUPED DATA

INTRODUCTION TO THE MEAN. Computing an average grade is perhaps a student's most familiar calculation. For example, if in four history quizzes Jennifer received 85, 90, 75, and 80, her average grade would be equal to (85 + 90 + 75 + 80) ÷ 4 = 330 ÷ 4 = 82.5

In other words, to get the average, we must first add all the numbers together. Then we divide this sum by the number of items that we added.

This calculation may be represented symbolically by the formula:

Arithmetic Average = ΣX/N. Where:

X represents an individual number

Σ is the Greek capital letter S and it is called sigma

Σ is a summation sign and always indicates addition

ΣX therefore, means the sum of all the individual numbers, i.e., the sum of all the X's

N represents the number of items which we added together to get ΣX, i.e., the number of X's

In our example above, we let $X_1 = 85$, $X_2 = 90$, $X_3 = 75$, and $X_4 = 80$. ΣX is equal to the sum of all the X's or: $X_1 + X_2 + X_3 + X_4$. ΣX is therefore equal to (85 + 90 + 75 + 80) = 330. N is equal to the number of X's or 4. The Arithmetic Average is:

$$= \frac{\Sigma X}{N} = \frac{330}{4} = 82.5$$

1. FORMULA FOR THE MEAN. In statistics the arithmetic average is called the "mean." It is represented by the symbol $\overline{X}$ Therefore, the formula that shows you how to find the mean is:

$$\overline{X} = \frac{\Sigma X}{N}$$

The mean is generally used to describe average: temperature, price, salary, grades, length of disease, height, weight, etc.

2. SAMPLE MEAN PROBLEMS.

EXAMPLE 1: The heights in inches of a five-man basketball team are 72, 70, 65, 68 and 75. Find the average mean height of this team:

Solution

Step 1. Mean = $\overline{X}$ = ΣX/N. To solve this equation we need to know ΣX and N.

Step 2. ΣX = sum of all the heights (X's) = 72 + 70 + 65 + 68 + 75 = 350"

Step 3. N = number of heights (X's) which we added together to get ΣX. N = 5

Step 4. $\overline{X} = \frac{\Sigma X}{N} = \frac{350}{5} = 70''$

EXAMPLE 2: The annual salaries of a seven-man accounting department are: $12,000; $8,000; $9,500; $7,000; $7,000; $7,600; and $7,000, respectively. What is the average mean salary of this group?

Solution

Step 1. Mean = $\overline{X}$ = ΣX/N. To solve this equation we need to know ΣX and N

Step 2. ΣX = sum of all the salaries (X's) = 12000 + 8000 + 9500 + 7000 + 7000 + 7600 + 7000 = $58,100

Step 3. $\overline{X} = \frac{\Sigma X}{N} = \frac{\$58100}{7} = \$8300$

EXAMPLE 3: Find the arithmetic mean of the numbers 10, 8, 7, 2 and 3:

Solution

Step 1. Mean = $\overline{X} = \frac{\Sigma X}{N}$ Needed : ΣX and N

Step 2. ΣX = Sum of all the numbers = 10 + 8 + 7 + 2 + 3 = 30

Step 3. N = number of items which we added to get ΣX. N = 5

Step 4. $\overline{X} = \frac{\Sigma X}{N} = \frac{30}{5} = 6$

EXAMPLE 4: Find the arithmetic mean of the numbers 6, 6, 6, 6, 6 and 6:

Solution

Step 1. $\overline{X} = \frac{\Sigma X}{N}$

Step 2. $\Sigma X = 6 + 6 + 6 + 6 + 6 + 6 = 36$

Step 3. $N = 6$

Step 4. $\overline{X} = \frac{\Sigma X}{N} = \frac{36}{6} = 6$

EXAMPLE 5: Find the arithmetic mean of the numbers 3, 6, 7, 0 and 4:

Solution

Step 1. $\overline{X} = \frac{\Sigma X}{N}$

Step 2. $\Sigma X = 3 + 6 + 7 + 0 + 4 = 20$

Step 3. $N = 5$ (remember 0 is a number, so we count it to get N)

Step 4. $\overline{X} = \frac{\Sigma X}{N} = \frac{20}{5} = 4$

EXAMPLE 6: Find the arithmetic mean of the numbers 45, 3, 9 and 7:

Solution

Step 1. $\overline{X} = \frac{\Sigma X}{N}$

Step 2. $\Sigma X = 45 + 3 + 9 + 7 = 64$

Step 3. $N = 4$

Step 4. $\overline{X} = \frac{\Sigma X}{N} = \frac{64}{4} = 16$

EXAMPLE 7: Find the arithmetic mean of the numbers 4, 0, 4, 4, 6, 7, 9, 10 and 9:

Solution

Step 1. $\overline{X} = \frac{\Sigma X}{N}$

Step 2. $\Sigma X = 4 + 0 + 4 + 4 + 6 + 7 + 9 + 10 + 9 = 53$

Step 3. $N = 9$

Step 4. $\overline{X} = \frac{\Sigma X}{N} = \frac{53}{9} = 5.89$

EXAMPLE 8: Find the arithmetic mean of the numbers 41, 52, 63, 34, 85 and 25:

Solution

Step l. $\overline{X} = \frac{\Sigma X}{N}$

Step 2. $\Sigma X = 41 + 52 + 63 + 34 + 85 + 25 = 300$

Step 3. $N = 6$

Step 4. $\overline{X} = \frac{\Sigma X}{N} = \frac{300}{6} = 50$

INTRODUCTION TO THE MEDIAN. The arithmetic average or mean can be influenced by a few extreme numbers in the data. The mean, therefore, may not always be typical of the data it represents. For example, suppose we entered a small law firm and asked the eleven employees what their average annual salaries are. The president earns $110,000. The other ten employees earn $5,000, $6,000, $6,000, $7,000, $7,500, $10,000, $11,000, $18,500, $19,000 and $20,000, respectively.

The average salary in this firm is thus equal to:

$$\overline{X} = \frac{110{,}000 + 5{,}000 + 6{,}000 + 6{,}000 + 7{,}000}{11}$$

$$\frac{+ 7{,}500 + 10{,}000 + 11{,}000 + 18{,}500 + 19{,}000 + 20{,}000}{}$$

$$= \frac{220{,}000}{11} = \$20{,}000$$

We can see from this example how one extreme value the $110,000 can influence the average. Realistically $20,000 in no way represents the salaries of this firm. In this example, an average called the "median" would have given us more information about this firm's salaries than the mean.

DEFINITION OF THE MEDIAN. When the data is arranged in the order of increasing or decreasing magnitude, the median is the point at which there are an equal number of items on either side. For an odd number of items, the median is the middle item. For an even number of items, the median is halfway between the middle two items. Thus, for an even number of items, we get the median by adding the middle two numbers together and then dividing their total by two.

To get the median salary in our example, we must first arrange the data in the order of magnitude: 5,000, 6,000, 6,000, 7,000, 7,500, 10,000, 11,000, 18,500, 19,000, 20,000 and 110,000. The median is the middle item (number) or 10,000. Note there are five numbers below 10,000 and five numbers above $10,000. We can see that $10,000 is a much more realistic average salary for this law firm than $20,000.

The median is generally used to describe average: life span for insurance purposes, size of families, amount of charitable contributions, income, etc.

1. SAMPLE MEDIAN PROBLEMS.

EXAMPLE 1: Find the median average for the numbers 2, 6, 8 and 4:

Solution

Step 1. Arrange the data in the order of magnitude: 2, 4, 6, 8

Step 2. The number of items (N) is even. Thus, the median is equal to the sum of the middle two numbers divided by 2

$$\text{Median} = \frac{(4+6)}{2} = \frac{10}{2} = 5$$

EXAMPLE 2: Find the median average for the numbers 7, 7, 2, 1, 2, 2, 1, 8 and 9:

Solution

Step 1. Arrange the data in the order of magnitude: 1, 1, 2, 2, 2, 7, 7, 8 and 9

Step 2. The number of items (N) is odd. Thus, the median is equal to the middle number. Median = 2

EXAMPLE 3: Find the median for the numbers 2, 4, 8, 8, 12 and 10:

Solution

Step 1. Arrange the data in the order of magnitude: 2, 4, 8, 8, 10, 12. The number of items (N) is even. Thus, the median is equal to the sum of the middle two numbers divided by 2

$$\text{Median} = \frac{(8+8)}{2} = \frac{16}{2} = 8$$

EXAMPLE 4: Find the median for the numbers 90, 0, 4, 7, 6 and 1:

Solution

Step 1. 0, 1, 4, 6, 7, 90

Step 2. $\text{Median} = \frac{(4+6)}{2} = \frac{10}{2} = 5$

EXAMPLE 5: Find the median for the numbers 5, 5, 0, 5 and 5:

Solution

Step 1. 0, 5, 5, 5, 5

Step 2. Median = 5

DEFINITION OF THE MODE. The mode is another form of average. It is defined as the number which occurs most frequently, i. e. the most popular number. The mode is generally used to describe average (most popular) clothes sizes, hair colors, etc.

1. SAMPLE "MODE" PROBLEMS.

EXAMPLE 1: Find the mode for the numbers 1, 2, 3, 5, 9, 7 and 8:

Solution

The mode = most popular number. In this case there is no mode.

EXAMPLE 2: Find the mode for the numbers 2, 2, 2, 4, 4, 4, 5, 6 and 7:

Solution

The mode = most popular number = 2 and 4 is called bimodal (two modes).

EXAMPLE 3: Find the mode for the numbers 2, 4, 6, 6, 8, 10, 12, 16, 18 and 20:

Solution

The mode = 6

EXAMPLE 4: In an English class 7 students have blue eyes, 2 have hazel eyes, and 3 have brown eyes. What is the modal eye color?

Solution

Mode = blue eyes

EXAMPLE 5: Last week a clothing store sold four size 36 suits, two size 42 suits, ten size 38 suits and one size 40 suit. What was the most popular (mode) suit size sold last week?

Solution

Mode = size 38

DEFINITION OF THE RANGE. Another item of information which we can get from our data is the "range." The range is equal to the smallest number subtracted from the largest number. Expressed symbolically:

$$\text{Range} = R = (X\ \text{max} - X\ \text{min})$$

The range indicates how far the data is spread out, since it is equal to the difference between the two extreme numbers (the largest and the smallest number).

1. SAMPLE RANGE PROBLEMS.

EXAMPLE 1: Find the range of the numbers 7, 5, 4, 1, and 2:

Solution

Range = (X max – X min)

= (7 – 1) = 6

EXAMPLE 2: Find the range of the numbers 20, 40, 600, 10, 0, 5 and 15:

Solution

Range = (X max – X min) = (600 – 0) = 600

EXAMPLE 3: Find the range of the numbers 8, 8, 8, 7, 7, 5, 5, 5, 2, 2 and 1:

Solution

R = (X max – X min) = (8 – 1) = 7

SAMPLE QUIZ:

Compute the mean, median, mode and range for the following problems:

EXAMPLE #20: 8, 6, 4, 12, 2, 10, 16, 14, 18, 20, 0

#21: 7, 7, 7, 1, 5, 9

#22: 15, 2, 15, 2, 15, 2, 15, 4, 4, 76

#23: 2.2, 2.4, 2.6, 2.6, 3.2

#24: 5, 5, 0, 5, 5

#25: 1, 2, 3, 4, 5

ANSWERS:

	#20	#21	#22	#23	#24	#25
Mean -	10	6	15	2.6	4	3
Median -	10	7	9.5	2.6	5	3
Mode -	none	7	15	2.6	5	none
Range -	20	8	74	1.0	5	4

2 STANDARD DEVIATION

HOW TO CALCULATE THE AVERAGE DEVIATION, VARIANCE AND STANDARD DEVIATION FROM UNGROUPED DATA

INTRODUCTION TO THE AVERAGE DEVIATION. If the heights in inches of a certain group of five people are 55, 60, 65, 60 and 70, respectively; we find the average mean height of this group by adding all the heights together and then dividing the total by 5.

$$\overline{X} = \frac{\Sigma X}{N} = \frac{55 + 60 + 65 + 60 + 70}{5} = \frac{310}{5} = 62"$$

Once we know what the average height is, we are usually interested next in comparing the individual heights to it. For example, if you are shorter than average you are usually interested in knowing exactly how much shorter than average you are.

The difference between an individual number and the average is called the deviation. The deviation or difference is equal to: $(X - \overline{X})$. The sum of all the deviations, $\Sigma(X - \overline{X})$, is always = 0

The deviation or differences in height are shown in Column 3 of the following table:

Height	Average Height	Deviation (difference) larger or smaller than average
X	$\overline{X}$	$X - \overline{X}$
55"	62"	−7" shorter than average
60"	62"	−2" shorter than average
60"	62"	−2" shorter than average
65"	62"	+3" taller than average
70"	62"	+8" taller than average

The sum of the deviation = 0. However, if we ignore the signs, we can find the average difference in the heights of this group by adding up all of the individual differences and then dividing the total by N. The differences in height are 7", 2", 2", 3", and 8", respectively. Thus, the average difference in height is equal to:

$$\frac{7" + 2" + 2" + 3" + 8"}{5} = \frac{22"}{5} = 4.4"$$

This means that on the average, the people in the group are either 4.4" shorter or 4.4" taller than the average height of 62"

1. DEFINITION OF AVERAGE DEVIATION. The average of the differences is called the average deviation and it is abbreviated A.D. The formula that shows you how to find the average deviation is:

$$\text{A.D.} = \frac{\Sigma|X - \overline{X}|}{N}$$

The notation | | means absolute value. Therefore, we ignore any minus signs inside this notation. For example, $|-7| = 7$, $|8 - 9| = |-1| = +1$, $|+6| = +6$. We use this notation because we are interested in the size of the differences and not in their sign.

Consequently, $\Sigma|X - \overline{X}|$ means the sum of all the absolute differences, i.e., the sum of all the $|X - \overline{X}|$'s. N is equal to the number of differences which we added together to get the $\Sigma|X - \overline{X}|$

2. SAMPLE AVERAGE DEVIATION PROBLEMS.

PROBLEM 1: Find the average deviation for the numbers 2, 5, 4, 9 and 5:

Solution

Step 1. $\text{A.D.} = \frac{\Sigma|X - \overline{X}|}{N}$

To solve this equation we need to know the values for N, $\overline{X}$, the $|X - \overline{X}|$'s

Step 2. $\overline{X} = \frac{\Sigma X}{N}$

Step 3. $\Sigma X = 2 + 5 + 4 + 9 + 5 = 25$

Step 4. $N = 5$

Step 5. $\overline{X} = \frac{\Sigma X}{N} = \frac{25}{5} = 5$

Step 6. Construct a table with the following column headings: Column 1 = X, Column 2 = $\overline{X}$, Column 3 = $X - \overline{X}$, Column 4 = $|X - \overline{X}|$:

Number	Average	Deviation	Absolute Deviation
X	$\overline{X}$	$X - \overline{X}$	$\lvert X - \overline{X}\rvert$

Step 7. Fill in the above table. Put the appropriate X's in Column 1 and the average $(\overline{X})$ in Column 2:

Number	Average	Deviation	Absolute Deviation
X	$\overline{X}$	$X - \overline{X}$	$\lvert X - \overline{X}\rvert$
2	5		
5	5		
4	5		
9	5		
5	5		

Step 8. Subtract the average $(\overline{X})$ in Column 2 from the individual numbers (X's) in Column 1 to get the individual $X - \overline{X}$'s in Column 3:

X	$\overline{X}$	$X - \overline{X}$	$\lvert X - \overline{X}\rvert$
2	5 →	−3	
5	5 →	0	
4	5 →	−1	
9	5 →	4	
5	5 →	0	

Step 9. Column 4 is equal to Column 3 without the minus signs:

X	$\overline{X}$	$X - \overline{X}$	$\lvert X - \overline{X}\rvert$
2	5	−3 →	3
5	5	0 →	0
4	5	−1 →	1
9	5	4 →	4
5	5	0 →	0

→ Arrow notation means work across the table.

Step 10. Add column 4 to get the sum of all the $|X - \overline{X}|$'s:

$$\Sigma|X - \overline{X}| = 3 + 0 + 1 + 4 + 0 = 8$$

Step 11. $\text{A.D.} = \dfrac{\Sigma|X - \overline{X}|}{N} = \dfrac{8}{5} = 1.6$

PROBLEM 2: Find the average deviation for the numbers 3, 5, 7 and 9:

Solution

Step 1. $\text{A.D.} = \dfrac{\Sigma|X - \overline{X}|}{N}$ (Needed: N, $\overline{X}$, the $|X - \overline{X}|$'s, and $\Sigma|X - \overline{X}|$)

Step 2. $\overline{X} = \dfrac{\Sigma X}{N}$

Step 3. $\Sigma X = 3 + 5 + 7 + 9 = 24$

Step 4. $N = 4$

Step 5. $\overline{X} = \dfrac{\Sigma X}{N} = \dfrac{24}{4} = 6$

Step 6. Construct the table and fill in the appropriate values:

Number	Average	Deviation	Absolute Deviation
X	$\overline{X}$	$X - \overline{X}$	$\lvert X - \overline{X}\rvert$
3	6 →	−3 →	3
5	6 →	−1 →	1
7	6 →	+1 →	1
9	6 →	+3 →	3

NOTE: The above table was filled out in one step to save space.

As a review:

Column 1 = individual number = X

Column 2 = average mean = $\overline{X}$

Column 3 = (Column 1 − Column 2) = $X - \overline{X}$

Column 4 = Column 3 without the minus signs = $|X - \overline{X}|$

NOTE: The sum of Column 3, $\Sigma | X - \overline{X} |$, always = 0

Step 7. Add Column 4 to get $\Sigma | X - | \overline{X}$

$$\Sigma|X - \overline{X}| = 3 + 1 + 1 + 3 = 8$$

Step 8. $\text{A.D.} = \dfrac{\Sigma|X - \overline{X}|}{N} = \dfrac{8}{4} = 2$

INTRODUCTION TO THE VARIANCE. In computation for the average deviation we ignored the minus signs in order to get the absolute difference between a number (X) and the average $(\overline{X})$. Instead of getting rid of the minus signs by ignoring them, we can accomplish the same thing by squaring the differences or deviations. The reason being that a minus number times a minus number is always equal to a plus number. Squaring the differences will therefore always give us a positive answer, for example:

$$-3^2 = (-3) \times (-3) = +9$$

1. DEFINITION OF THE VARIANCE. We can compute the average of the squared deviations by first squaring the individual deviations or differences, then adding them, and finally dividing the total by N. The average of the squared deviations is called the variance and is represented by the letter S. Therefore, the formula that shows you how to find the variance is:

$$S = \frac{\Sigma(X - \overline{X})^2}{N}$$

2. SAMPLE VARIANCE PROBLEMS.

PROBLEM 1: Find the variance of the numbers 1, 2, 5, 4 and 8:

Solution

Step 1. $\text{Variance} = S = \frac{\Sigma(X - \overline{X})^2}{N}$

To solve this equation we need to know the values for N, $\overline{X}$, plus the

$(X - \overline{X})$'s, the $(X - \overline{X})^2$'s and $\Sigma(X - \overline{X})^2$

Step 2. $\overline{X} = \frac{\Sigma X}{N} = \frac{+1+2+5+4+8}{5} = \frac{20}{5} = 4$

Step 3. The easiest way of finding the values needed to solve the equation for the variance is by constructing a table similar to the one used to solve the average deviation formula:

Number	Average	Deviation	Deviation Squared
X	$\overline{X}$	$X - \overline{X}$	$(X - \overline{X})^2$
1	4	−3 →	9
2	4	−2 →	4
5	4	+1 →	1
4	4	0 →	0
8	4	+4 →	16

NOTE: To save space the above table was filled out in one step.

Column 1 = individual number = X

Column 2 = average mean = $\overline{X}$

Column 3 = (individual X in Column l minus $\overline{X}$ in Column 2) $= (X - \overline{X})$

Column 4 = Column 3 squared = $(X - \overline{X})^2$

Step 4. Add the individual $(X - \overline{X})^2$'s in Column 4 to get:

$$\Sigma(X - \overline{X})^2$$

$$\Sigma(X - \overline{X})^2 = 9 + 1 + 4 + 16 = 30$$

Step 5. $S = \frac{\Sigma(X - \overline{X})^2}{N} = \frac{30}{5} = 6$

INTRODUCTION TO THE STANDARD DEVIATION. In the derivation of the formula for the variance from the average deviation, we squared the individual deviations (differences) in order to eliminate the minus signs. If we now undo this squaring by taking the square root of the equation for the variance, we get the expression:

$$\sqrt{\frac{\Sigma(X - \overline{X})^2}{N}}$$

This expression is called the standard deviation and it is represented by the symbol σ.

1. DEFINITION OF STANDARD DEVIATION. σ is the small Greek letter s and is called sigma. In statistics the symbol σ has no relation to the symbol Σ, even though they are both called sigma. The formula that shows you how to find the standard deviation is:

$$\sigma = \sqrt{\frac{\Sigma(X - \overline{X})^2}{N}}$$

NOTE: The formula for the standard deviation was derived from the formula for the average deviation. It is not exactly equal to the average deviation, because of the squaring and taking the square root. The standard deviation has no verbal definition. It is only defined by its formula. It may help the student to visualize the concept of the standard deviation by looking at it as a way of measuring the average differences or variability between the numbers in the data.

2. SAMPLE STANDARD DEVIATION PROBLEMS. The following four sample problems have the same average mean, but different variations in their data. The standard deviations, therefore, will tell us how the numbers vary from the mean of the data. A relatively large σ will indicate large differences between the numbers in the data, and conversely, a relatively small σ will indicate small differences between the numbers.

PROBLEM 1: Compute the standard deviation for the numbers 50, 60 and 70:

Solution

Step 1. Standard deviation = $\sigma = \sqrt{\frac{\Sigma(X - \overline{X})^2}{N}}$

Needed: N, $\overline{X}$, the $(X - \overline{X})$'s, the $(X - \overline{X})^2$'s and $\Sigma(X - \overline{X})^2$.

Step 2. $\overline{X} = \frac{\Sigma X}{N}$

$= \frac{50 + 60 + 70}{3} = \frac{180}{3} = 60$

Step 3. The easiest way of finding values needed to solve the equation for the standard deviation is by constructing a table similar to the ones used to solve the average deviation and variance formulas:

Number	Average	Deviation	Deviation squared
X	$\overline{X}$	$X - \overline{X}$	$(X - \overline{X})^2$
50	60	−10 →	100
60	60	0 →	0
70	60	+10 →	100

NOTE: To save space the above table was filled out in one step:

Column 1 = individual number = X

Column 2 = average mean = $\overline{X}$

Column 3 = (individual X in Column l minus $\overline{X}$ in Column 2) $= (X - \overline{X})$.

Column 4 = Column 3 squared = $(X - \overline{X})^2$

NOTE: The sum of Column 3, the $\Sigma(X - \overline{X})$ always = 0

Step 4. Add the individual $(X - \overline{X})^2$'s in Column 4 to get $\Sigma(X - \overline{X})^2$

$\Sigma(X - \overline{X})^2 = 100 + 0 + 100$

$= 200$

Step 5. $\sigma = \sqrt{\frac{\Sigma(X - \overline{X})^2}{N}}$

$= \sqrt{\frac{200}{3}} = \sqrt{66.7} = 8.2$

PROBLEM 2: Compute the standard deviation for the numbers 59, 60, 61:

Solution

Step 1. Standard deviation = $\sigma = \sqrt{\frac{\Sigma(X - \overline{X})^2}{N}}$

Step 2. $\overline{X} = \frac{\Sigma X}{N}$

$= \frac{59 + 60 + 61}{3} = \frac{180}{3} = 60$

Step 3. Construct the table:

Number	Average	Deviation	Deviation squared
X	$\overline{X}$	$X - \overline{X}$	$(X - \overline{X})^2$
59	60	−1 →	1
60	60	0 →	0
61	60	+1 →	1

Column 1 = individual number = X

Column 2 = average mean = $\overline{X}$

Column 3 = (individual X in Column 1 minus $\overline{X}$ in Column 2) $= (X - \overline{X})$

Column 4 = Column 3 squared = $(X - \overline{X})^2$

CHECK: $\Sigma(X - \overline{X}) = -1 + 0 + 1 = 0$

Step 4. Add the individual $(X - \overline{X})^2$'s in Column 4 to get $\Sigma(X - \overline{X})^2$

$\Sigma(X - \overline{X})^2 = 1 + 0 + 1 = 2$

Step 5. $\sigma = \sqrt{\frac{\Sigma(X - \overline{X})^2}{N}} = \sqrt{\frac{2}{3}} = \sqrt{.667} = .82$

The differences in the data are less in this problem than in the previous problem. Therefore, σ is smaller, as you can see.

PROBLEM 3: Compute the standard deviation for the numbers 60, 60 and 60:

Solution

Step 1. Standard deviation $\sigma = \sqrt{\frac{\Sigma(X - \overline{X})^2}{N}}$

Needed: N, $\overline{X}$, the $(X - \overline{X})$'s, the $(X - \overline{X})^2$'s, and $\Sigma(X - \overline{X})^2$

Step 2. $\overline{X} = \frac{\Sigma X}{N} = \frac{60+60+60}{3} = \frac{180}{3} = 60$

Step 3. Construct the table:

Number	Average	Deviation	Deviation squared
X	$\overline{X}$	$X-\overline{X}$	$(X-\overline{X})^2$
60	60	0	0
60	60	0	0
60	60	0	0

Column 1 = individual number = X

Column 2 = average mean = $\overline{X}$

Column 3 = (individual X in Column l minus $\overline{X}$ in Column 2) $= (X-\overline{X})$

Column 4 = Column 3 squared = $(X-\overline{X})^2$

Step 4. Add all the individual $(X-\overline{X})^2$'s in Column 4 to get:

$\Sigma(X-\overline{X})^2$

$= 0 + 0 + 0 = 0.$

Step 5. $\sigma = \sqrt{\frac{\Sigma(X-\overline{X})^2}{N}} = \sqrt{\frac{0}{3}} = 0$

There were no differences among the numbers in our data so σ came out to be equal to zero.

PROBLEM 4: Compute the mean and standard deviation for the numbers 3, 7, 10, 6 and 4:

Solution

Step 1. $\overline{X} = \frac{\Sigma X}{N}$

Needed: N, $\overline{X}$, the $(X-\overline{X})$'s, the $(X-\overline{X})^2$'s and $\Sigma(X-\overline{X})^2$

Step 2. $\overline{X} = \frac{\Sigma X}{N} = \frac{3+7+10+6+4}{5} = \frac{30}{5} = 6$

Step 3. Construct the table:

X	$\overline{X}$	$X-\overline{X}$	$(X-\overline{X})^2$
3	6	−3 →	9
7	6	1 →	1
10	6	+4 →	16
6	6	0 →	0
4	6	−2 →	4

Column 3 = Individual X in Column 1 minus $\overline{X}$ in Column 2

Column 4 = Column 3 squared = $(X-\overline{X})^2$

Step 4. $\Sigma(X-\overline{X})^2 = 9 + 1 + 16 + 0 + 4 = 30$

CHECK: $\Sigma(X-\overline{X}) = -3 + 1 + 4 + 0 - 2 = 0$

Step 5. $\sigma = \sqrt{\frac{\Sigma(X-\overline{X})^2}{N}} = \sqrt{\frac{30}{5}} = \sqrt{6} = 2.45$

SAMPLE QUIZ:

Compute the mean and the standard deviation for the following problems:

EXAMPLE:

#8:	5	5	4	6	
#9:	10	8	7	3	2
#10:	3	6	7	0	4
#11:	5	3	9	7	

ANSWERS:

	#8	#9	#10	#11
$\overline{X}$ =	5	6	4	6
σ =	$\sqrt{.5}$	$\sqrt{9.2}$	$\sqrt{6}$	$\sqrt{2.24}$
	.707	3.03	2.45	1.5

3 FREQUENCY

HOW TO USE FREQUENCIES IN COMPUTING THE MEAN AND STANDARD DEVIATION FROM UNGROUPED DATA

INTRODUCTION TO FREQUENCY. Multiplication is actually a quick way of doing addition. For example, 5 × 3 means take 5 three times, that is, 5 + 5 + 5. We know that 5 + 5 + 5 = 15 and from our multiplication tables 5 × 3 = 15, so our statement checks.

EXAMPLES:

(a) 6 × 3 = 6 + 6 + 6 = 18

(b) 2 × 2 = 2 + 2 = 4

(c) 6 × 1 = 6

(d) 8 + 8 + 8 + 8+ 8 = 8 × 5 = 40

(e) 9 + 9 = 9 × 2 = 18

(f) 7 = 7 × 1 = 7

1. DEFINITION OF FREQUENCY. The term frequency can be defined as the number of times that a given item occurs, i.e., how often. Frequency is represented by the letter F.

   The frequencies in the above examples are:

   (a) F = 3 (there are three 6's)

   (b) F = 2 (there are two 2's)

   (c) F = 1 (there is one 6)

   (d) F = 5 (there are five 8's)

   (e) F = 2 (there are two 9's)

   (f) F = 1 (there is only one 7)

2. FREQUENCY AND THE MEAN. If we wanted to find the arithmetic mean for the numbers 2, 2, 2, 2, 4, 4, 4, 6, 6 and 8, we would simply add them together and then divide by 10 (there are ten numbers):

$$\overline{X} = \frac{2+2+2+2+4+4+4+6+6+8}{10}$$

$$= \frac{40}{10} = 4$$

Since we have four 2's, three 4's, two 6's and one 8, we can also compute the arithmetic mean as follows:

$$\overline{X} = \frac{(4\times 2)+(3\times 4)+(2\times 6)+(1\times 8)}{10}$$

$$= \frac{8+12+12+8}{10} = \frac{40}{10} = 4$$

This computation can be represented by the Formula;

$$\overline{X} = \frac{\Sigma FX}{N}$$

ΣFX means the sum of all the FX's, i.e., the first number multiplied by its frequency, plus the second number multiplied by its frequency, plus the third number multiplied by its frequency, and so on. Expressed symbolically:

$$\Sigma FX = [(F_1)(X_1) + (F_2)(X_2) + (F_3)(X_3) + (F_4)(X_4) + \text{etc.}]$$

N is equal to the number of individual items which we added together to get ΣFX. N therefore is equal to ΣF. Thus, $N = (F_1 + F_2 + F_3 + F_4 + \text{etc.})$.

In our problem above:

$F_1 = 4, X_1 = 2, F_2 = 3, X_2 = 4, F_3 = 2, X_3 = 6, F_4 = 1, X_4 = 8$

$$\Sigma FX = (4 \times 2) + (3 \times 4) + (2 \times 6) + (1 \times 8)$$
$$= 8 + 12 + 12 + 8 = 40$$

$$N = \Sigma F = 4 + 3 + 2 + 1 = 10$$

$$\overline{X} = \frac{\Sigma FX}{N} = \frac{40}{10} = 4$$

3. SAMPLE MEAN PROBLEMS:

PROBLEM 1: Find the arithmetic mean of the numbers 9, 8, 7, 6, 7, 8, 9, 6, 9, 6, 8, 7, 9, 6, 8, 9, 7, 7, 6 and 8:

Solution (A):

$$\overline{X} = \frac{\Sigma X}{N} = \frac{9+8+7+6+7+8+9+6+9}{20}$$

$$\frac{+6+8+7+9+6+8+9+7+7+6+8}{} = \frac{150}{20}$$

$$\overline{X} = \frac{150}{20} = 7.5$$

Solution (B):

In our data we have five 9's, five 8's, five 7's, and five 6's:

$$\overline{X} = \frac{\Sigma FX}{N} = \frac{(5\times9)+(5\times8)+(5\times7)+(5\times6)}{5+5+5+5}$$

$$= \frac{45+40+35+30}{20} = \frac{150}{20} = 7.5$$

Solution (C):

Step 1. To simplify our calculation, we can arrange our data in a table having the following column headings: Column 1 = X, Column 2 = F, and Column 3 = FX:

Number	Frequency	Number Times Its Frequency
X	F	FX

Step 2. Fill in the above table. Put the appropriate X's in Column 1 and corresponding F's in Column 2:

Number	Frequency	Number Times Its Frequency
X	F	FX
9	5	
8	5	
7	5	
6	5	

Step 3. Multiplying across, multiply the X in Column 1 by its corresponding F in Column 2 to get FX in Column 3:

Number	Frequency	Number Times Its Frequency
X	F	FX
9	5 →	45
8	5 →	40
7	5 →	35
6	5 →	30

Step 4. Add up Column 3 to get $\Sigma FX = 45 + 40 + 35 + 30 = 150$

Step 5. Add up Column 2 to get $N = \Sigma F = 5 + 5 + 5 + 5 = 20$

Step 6. $\overline{X} = \frac{\Sigma FX}{N} = \frac{150}{20} = 7.5$

PROBLEM 2: Find the arithmetic mean for ten 8's, four 40's, two 30's, and five 2's:

Solution

Step 1. $\overline{X} = \frac{\Sigma X}{N} = \frac{\Sigma FX}{N}$. Needed: N, the FX's and FX

Step 2. Construct the table:

X	F	FX
8	10 →	80
40	4 →	160
30	2 →	60
2	5 →	10

Column 3 = individual number (X) in Column 2 times its corresponding frequency in Column 2

Step 3. Add up Column 3 to get $\Sigma FX = 80 + 160 + 60 + 10 = 310$

Step 4. Add up Column 2 to get $N = \Sigma F = 10 + 4 + 2 + 5 = 21$

Step 5. $\overline{X} = \frac{\Sigma FX}{N} = \frac{310}{21} = 14.76$

4. FREQUENCY AND STANDARD DEVIATION: If we wanted to find the standard deviation for a group of numbers, we would: first find $\overline{X}$, then subtract $\overline{X}$ from all the individual numbers to get the $(X-\overline{X})$'s, next square all the differences to get the $\Sigma(X-\overline{X})^2$'s, next divide the sum of the squares of the individual differences by N to get $\frac{?(X-\overline{X})^2}{N}$, and finally take the square root of:

$$\frac{\Sigma(X-\overline{X})^2}{N} \text{ to get } \sigma = \sqrt{\frac{(X-\overline{X})^2}{N}}$$

For example, find the standard deviation for the numbers 2, 2, 2, 2, 4, 4, 4, 6, 6 and 8. This is the same problem that we worked out in the section "Frequency and the Mean." To solve the standard deviation formula, we need the following values: N, $\overline{X}$, and the $(X-\overline{X})$'s, the $(X-\overline{X})^2$'s and $\Sigma(X-\overline{X})^2$;

$$X = \frac{\Sigma X}{N} = \frac{2+2+2+2+4+4+4+6+6+8}{10}$$

$$= \frac{40}{10} = 4$$

Next we construct a table in order to find the values needed to solve the formula:

X	$\overline{X}$	$X-\overline{X}$	$(X-\overline{X})^2$
2	4	–2 →	4
2	4	–2 →	4
2	4	–2 →	4
2	4	–2 →	4
4	4	0 →	0
4	4	0 →	0
4	4	0 →	0
6	4	+2 →	4
6	4	+2 →	4
8	4	+4 →	16

$\Sigma(X-\overline{X})^2$ = sum of column four = 4 + 4 + 4 + 4 + 0 + 0 + 0 + 4 + 4+16 = 40X;

$$\sigma = \sqrt{\frac{\Sigma(X-\overline{X})^2}{N}} = \sqrt{\frac{40}{10}} = \sqrt{4} = 2$$

We can see that in Column 1, of the above table, we have four 2's, three 4's, two 6's, and one 8. In Column 3, we have four (2 – 4)'s, three (4 – 4)'s, two (6 – 4)'s and one (8 – 4). The four, three, two and one correspond to the frequencies of the numbers in our data, namely, 2, 4, 6, and 8 respectively.

Therefore, if we use the frequencies in computing σ, we find that the formula for the standard deviation is also equal to:

$$\sqrt{\frac{\Sigma F(X-\overline{X})^2}{N}}$$

$\Sigma F(X-\overline{X})^2$ means

$$F_1(X_1-\overline{X})^2 + F_2(X_2-\overline{X})^2 + F_3(X_3-\overline{X})^2 + \text{ etc.}$$

Our table headings for computing the standard deviation will also change accordingly.

If we insert the data from the above problem into the new table, we would get the following:

X	F	$\overline{X}$	$(X-\overline{X})$	$(X-\overline{X})^2$	$F(X-\overline{X})^2$
2	4	4	–2 →	4	16
4	3	4	0 →	0	0
6	2	4	2 →	4	8
8	1	4	4 →	16	16

Column 1 = individual number = X

Column 2 = corresponding frequency = F

Column 3 = average mean = $\overline{X}$

Column 4 = (Column 1 minus Column 3)
= $X-\overline{X}$ (subtract across)

Column 5 = Column 4 squared = $(X-\overline{X})^2$

Column 6 = (Column 5 multiplied by Column 2)

= $F(X-\overline{X})^2$ (multiply across)

Sum of Column 6 = $\Sigma F(X-\overline{X})^2 = 40$

Since the formula for computing the standard deviation from frequency data is equal to:

$$\sigma = \sqrt{\frac{\Sigma F(X-\overline{X})^2}{N}}$$

$$\sigma = \sqrt{\frac{40}{10}} = \sqrt{4} = 2$$

NOTE: This is the same answer as the one we got before when we didn't use the frequencies. However, by using the frequencies we were able to simplify our arithmetic considerably. You can also see that because of the frequencies, the sum of the $(X-\overline{X})$ column is no longer = 0. However, now $(X-\overline{X}) = \sigma$

5. SAMPLE STANDARD DEVIATION PROBLEMS.

PROBLEM 1: Find the standard deviation for the numbers 5, 5, 3, 2, 6, 6, 6, 8, 8, 8, 5, 8, 2, 2, 2, 2, 2, 2, 2, 2, 2, 2, 2, 8 and 25

Solution:

Step 1. An examination of our data shows that we have 12 two's, 3 five's, 5 eight's, 3 six's, 1 three and 1 twenty-five. Since there are 25 numbers we can simplify our calculations by using the frequency method of calculating σ. We would therefore have only 6 numbers to square and add instead of 25:

Step 2. $\sigma = \sqrt{\frac{\Sigma(X-\overline{X})^2}{N}} = \sqrt{\frac{\Sigma F(X-\overline{X})^2}{N}}$

Needed: N, $\overline{X}$, the $(X-\overline{X})$'s, the $(X-\overline{X})^2$'s, the $F(X-\overline{X})^2$'s, and the $\Sigma F(X-\overline{X})^2$

Step 3. Since we are going to use frequencies to calculate σ, we might just as well use them to calculate $\overline{X}$ too. So, to simplify our calculations, we are going to combine the tables for $\overline{X}$ and for σ into one table:

X	F	FX	$\overline{X}$	$X-\overline{X}$	$(X-\overline{X})^2$	$F(X-\overline{X})^2$
2	12	24	5	−3 →	9	108
3	1	3	5	−2 →	4	4
5	3	15	5	0 →	0	0
6	3	18	5	+1 →	1	3
8	5	40	5	+3 →	9	45
25	1	25	5	+20 →	400	400

Step 4. Multiplying across, Column 3 (FX) = (Column 1 (X) times Column 2 (F))

Step 5. $\overline{X} = \frac{\Sigma FX}{N} = \frac{\text{Sum of Column 3}}{\text{Sum of Column 2}}$

$= \frac{24+3+15+18+40+25}{12+1+3+3+5+1} = \frac{125}{25} = 5$

Step 6. Insert $\overline{X}$ = 5 in Column 4 in the table above

Step 7. Subtracting across, Column 5 $(X-\overline{X})$ = (Column 1 (X) minus Column 4 ($\overline{X}$))

Step 8. Column 6 = Column 5 squared = $(X-\overline{X})^2$

Step 9. Column 7 = $F(X-\overline{X})^2$ = (Column 6 times Column 2). (Multiply across)

Step 10. $\Sigma F(X-\overline{X})^2$ = sum of Column 7 = (108 + 4 + 0 +3+ 45 + 400) = 560

Step 11. $\sigma = \sqrt{\frac{\Sigma F(X-\overline{X})^2}{N}} = \sqrt{\frac{560}{25}} = \sqrt{22.4} = 4.73$

PROBLEM 2: Find the standard deviation for the numbers 4, 4, 4, 4, 4, 4, 4, 4, 4, 4,5, 5, 5, 5, 5, 7, 7, 7, 7 and 7:

Solution

Step 1. Since we have 10 fours, 5 fives, and 5 sevens we can simplify our calculations by using the frequency method of computing σ

Step 2. We construct our table to get the needed values:

X	F	FX	$\overline{X}$	$(X-\overline{X})$	$(X-\overline{X})^2$	$F(X-\overline{X})^2$
4	10	40	5	−1 →	1 →	10
5	5	25	5	0 →	0 →	0
7	5	35	5	+2 →	4 →	20

Step 3. From our data, we enter the appropriate items in Columns 1 and 2

Step 4. Column 3 = (Column 2 times Column 1)

FX = F times X

Step 5. $\overline{X} = \frac{\Sigma FX}{N} = \frac{\text{Sum of Column 3}}{\text{Sum of Column 2}}$

$= \frac{40+25+35}{10+5+5} = \frac{100}{20} = 5$

Step 6. Enter $\overline{X}$ = 5 in Column 4 of the above table

Step 7. Column 5 = Column 1 minus Column 4;

$(X-\overline{X}) = X - \overline{X}$

Step 8. Column 6 = Column 5 squared.

$(X-\overline{X})^2 = (X-\overline{X})(X-\overline{X})$

Step 9. Column 7 = Column 6 times Column 2.

$F(X-\overline{X})^2 = (X-\overline{X})^2 F$

Step 10. $\Sigma F(X-\overline{X})^2$ = sum of Column 7 = 10 + 0 + 20 = 30

Step 11. $\sigma = \sqrt{\frac{\Sigma F(X-\overline{X})^2}{N}} = \sqrt{\frac{30}{20}} = \sqrt{1.5} = 1.22$

PROBLEM 3: Compute the mean and standard deviation for the following numbers:

70	70	70	70	60
60	60	50	50	50
50	50	50	50	50
15	15	80	80	80
80	80	50	80	80

Solution

Step 1. Since we have two 15's, nine 50's, three 60's, four 70's and seven 80's, we can simplify our calculations by using the frequency method of computing the mean and standard deviation

Step 2. We construct our table to get the needed values:

X	F	FX	$\overline{X}$	$X-\overline{X}$	$(X-\overline{X})^2$	$F(X-\overline{X})^2$
15	2	30	60	−45 →	2025 →	4050
50	9	450	60	−10 →	100 →	900
60	3	180	60	0 →	0 →	0
70	4	280	60	+10 →	100 →	400
80	7	560	60	+20 →	400 →	2800

Step 3. From our data we enter the appropriate items in Column 1 and 2

Step 4. Column 3 = (Column 2 times Column 1): FX = F times X. (Multiplying across)

Step 5. $\overline{X} = \frac{\Sigma FX}{N} = \frac{\text{Sum of Column 3}}{\text{Sum of Column 2}}$

$$= \frac{30 + 450 + 180 + 280 + 550}{2 + 9 + 3 + 4 + 7} = \frac{1500}{25}$$

$\overline{X} = 60$

Step 6. Enter $\overline{X} = 60$ in Column 4 of the above table.

Step 7. Column 5 = (Column 1 minus Column 4): $(X - \overline{X}) = X - \overline{X}$. (Work across)

Step 8. Column 6 = Column 5 squared, i.e., Column 5 multiplied by itself. $(X - \overline{X})^2 = (X - \overline{X})(X - \overline{X})$. (Work across)

Step 9. Column 7 = (Column 6 times Column 2).

$F(X - \overline{X})^2 = (X - \overline{X})^2(F)$. (Work across)

Step 10. $\Sigma F(X - \overline{X})^2$ = sum of Column 7. (4050 + 900 + 0 + 400 + 2800) = 8150

Step 11. $\sigma = \sqrt{\frac{\Sigma F(X - \overline{X})^2}{N}} = \sqrt{\frac{8150}{25}} = \sqrt{326} = 18.06$

SAMPLE QUIZ:

PROBLEM 1: Use the frequency method to compute the mean of the numbers 130,88,103, 66, 88, 88, 77, 66, 77 and 77:

Answer: $\overline{X} = 86$

PROBLEM 2: Use the frequency method to compute the mean of the numbers 0, 4, 5, 7, 9, 10, 12, 26, 26, 12, 10, 9, 5, 0, 0, 9, 9, 9, 12 and 26:

Answer: $\overline{X} = 10$

PROBLEM 3: Use the frequency method to compute the standard deviation of the numbers 2, 2, 4, 4, 6, 6, 8, 8, 2, 2, 8, 2, 2, 2 and 2:

Answer: $\overline{X} = 4$, $\sigma = \sqrt{\frac{88}{15}} = \sqrt{5.87} = 2.42$

4 ASSUMED AVERAGE

HOW TO USE AN ASSUMED AVERAGE IN COMPUTING THE MEAN FROM UNGROUPED DATA

INTRODUCTION TO THE ASSUMED AVERAGE. If we subtract a number from both sides of an equation, it has no effect on the equality, since we are actually reducing both sides by an equal amount. That is, after the subtraction both sides will still be equal to each other. For example, in the equation Y = 4X, if X = 2 then Y = 4(2) = 8. In other words 8 = 4(2) = 8 or 8 = 8. Now if we subtract 3 from both sides of this equation, the equation becomes: Y − 3 = (4X) − 3. Substituting X = 2 and Y = 8, we get 8 − 3 = 4(2) − 3 or 5 = 5

If we subtract a number "A" from both sides of the equation for the mean, we would get the expression:

$$\overline{X} - A = \frac{\Sigma X}{N} - A, \text{ combining terms, this equation becomes}$$

$$\overline{X} = A + \frac{\Sigma(X - A)}{N}. \text{ When } A = 0, \text{ then}$$

$$X = 0 + \frac{\Sigma X - 0}{N} = \frac{\Sigma X}{N} \text{ (our original equation)}$$

If we use frequencies, our equation for the mean becomes equal to:

$$\overline{X} = A + \frac{\Sigma F(X - A)}{N}$$

Now if we let "A" represent an assumed number or assumed average, then the formula that shows you how to find the mean by using an assumed average is:

$$\overline{X} = A + \frac{\Sigma(X - A)}{N} \text{ and } \overline{X} = A + \frac{\Sigma F(X - A)}{N}$$

The assumed average method of computing $\overline{X}$ allows you to reduce the size of the numbers that you have to work with. You can assume any number you want. However, its usually best to assume either the number with the highest frequency, or if you are not using frequencies, let "A" equal the largest number in the data.

PROBLEM 1: Use an assumed average to compute the mean for the numbers 2, 6, 7, 1 and 4:

Solution (a):

Step l: $\overline{X} = A + \frac{\Sigma(X - A)}{N}$ Needed: N, (X − A) and Σ(X − A)

Step 2. Let's assume A = 6

Step 3.

$$\overline{X} = 6 + \frac{(2-6)+(6-6)+(7-6)+(1-6)+(4-6)}{5}$$

$$\overline{X} = 6 + \frac{-4+0+1-5-2}{5}$$

$$\overline{X} = 6 + \frac{-10}{5}$$

$$\overline{X} = 6 - 2 = 4$$

Solution (b):

Assume A = 15

$$\overline{X} = 15 + \left(\frac{(2-15)+(6-15)+(7-15)+(1-15)+(4-15)}{5}\right)$$

$$\overline{X} = 15 + \left(\frac{-13-9-8-14-11}{5}\right)$$

$$\overline{X} = 15 + \left(\frac{-55}{5}\right) = 15 - 11 = 4$$

Solution (c):

Assume A = 1

$$\overline{X} = 1 + \left(\frac{(2-1)+(6-1)+(7-1)+(1-1)+(4-1)}{5}\right)$$

$$\overline{X} = 1 + \left(\frac{+1+5+6+0+3}{5}\right)$$

$$\overline{X} = 1 + \left(\frac{15}{5}\right)$$

$$\overline{X} = 1 + 3$$

$$\overline{X} = 4$$

Solution (d):

Assume A = 0 (regular method of computing $\overline{X}$:

$$\overline{X} = 0 + \frac{(2-0)+(6-0)+(7-0)+(1-0)+(4-0)}{5}$$

$$\overline{X} = 0 + \frac{2+6+7+1+4}{5}$$

$$\overline{X} = 0 + \frac{20}{5}$$

$$\overline{X} = 0 + 4$$

$$\overline{X} = 4$$

PROBLEM 2: Use an assumed average to compute the mean for the numbers 502, 506, 507, 501 and 504:

Solution

Step l. $\overline{X} = A + \frac{\Sigma(X-A)}{N}$

Needed: N, (X – A), Σ(X – A)

Step 2. Assume A = 502 (Remember, we can assume any number we want to):

Step 3. $$\overline{X} = 502 + \left(\frac{(502-502)+(506-502)+(507-502)+(501-502)+(504-502)}{5}\right)$$

$$\overline{X} = 502 + \left(\frac{0+4+5-1+2}{5}\right)$$

$$\overline{X} = 502 + (2)$$

$$\overline{X} = 504$$

By assuming A = 502, we only had to add one digit numbers together to get $\overline{X}$, thus simplifying our calculations.

PROBLEM 3: Use an assumed average to compute the mean for the numbers 3, 3, 3, 7, 7, 9, 9, 9, 9, 10, 10, 5, 5, 1 and 0:

Solution

Step 1. We have three 3's, two 7's, four 9's, two 10's, two 5's, one 1 and one 0. Therefore, instead of working with 15 numbers, we can use the frequency method and end up working with only 7 numbers.

$$\overline{X} = A + \frac{\Sigma F(X-A)}{N}$$

Needed: N, (X – A)'s, F(X – A)'s and ΣF(X – A).\

Step 2. Assume A = 9 (number with the highest frequency):

Step 3. $$\overline{X} = 9 + \left(\frac{+3(3-9)+2(7-9)+4(9-9)+2(10-9)+2(5-9)+1(1-9)+1(0-9)}{5}\right)$$

$$\overline{X} = 9 + \left(\frac{+3(-6)+2(-2)+4(0)+2(1)+2(-4)+1(-8)+1(-9)}{15}\right)$$

$$\overline{X} = 9 + \left(\frac{-18-4+0+2-8-8-9}{15}\right)$$

$$\overline{X} = 9 + \left(\frac{-45}{15}\right)$$

$$\overline{X} = 9 - 3$$

$$\overline{X} = 6$$

PROBLEM 4: Use an assumed average to compute the mean for the numbers 10,20, 30 and 40:

Solution

Step 1. $\overline{X} = A + \frac{\Sigma(X-A)}{N}$

Step 2. Let's assume A = 100 to show that any number can be assumed.

Step 3.

$$\overline{X} = 100 + \left(\frac{(10-100)+(20-100)+(30-100)+(40-100)}{4}\right)$$

$$\overline{X} = 100 + \left(\frac{-90-80-70-60}{4}\right)$$

$$\overline{X} = 100 + \left(\frac{-300}{4}\right)$$

$$\overline{X} = 100 - 75$$

$$\overline{X} = 25$$

5 GROUPED DATA

HOW TO COMPUTE THE MEAN, MEDIAN, MODE, RANGE, AND STANDARD DEVIATION FROM GROUPED DATA

INTRODUCTION TO CLASS INTERVALS. So far, the problems which we worked on involved a small amount of data—usually less than ten numbers. In actual practice, however, we generally work with problems that involve a great deal of data. In these situations, the methods which we learned for computing $\overline{X}$ and σ could prove to be very laborious and time consuming. For example, just imagine how long it would take you to compute the standard deviation for 5,000 three-digit numbers.

For most practical purposes, exact numbers or values are not really needed. So, in order to condense our data into a manageable form, we can let one number represent a group of numbers which are close to it in value.

Perhaps the easiest way of illustrating data grouping is with a problem which should be familiar to most students.

EXAMPLE: Suppose the following numbers represent the average mean final grades which 25 students received in an English class:

99	92	83	94	92
79	91	85	76	77
74	55	85	70	60
83	85	68	65	63
68	87	57	50	62

If an A represents a grade in the 90's, a B represents a grade in the 80's, a C represents a grade in the 70's, and so on; find the number of A's, B's, C's, D's, and E's in this class:

Solution: We are going to condense our 25 individual grades into five groups each of which is represented by a letter grade. This can best be accomplished by using a table as shown below:

Letter Grade	Range of Grades Represented by the Letter Grades	Individual Grades Represented by the Letter Grade	Number of Students Receiving the Letter Grade or Frequency
A	90 to 99	99,92,91,94,92	5
B	80 to 89	83,85,87,83,85,85	6
C	70 to 79	79,74,76,70,77	5
D	60 to 69	68,68,65, 60,63,62	6
E	50 to 59	55,57, 50	3

Thus, the students in this English class received five A's, six B's, five C's, six D's, and three E's.

Instead of letting a letter grade represent a certain group or range of grades, we can also represent the group by a number grade.

For example, we can say that the letter grade "A" is equivalent to a 95, the letter grade "B" is equivalent to a 85, "C" is equivalent to a 75, and so on. The number grades which we, in turn, picked to represent the letter grades are actually the midpoints of the range of grades represented by the letter grade. That is, 95 is halfway between 90 and 99, 85 is halfway between 80 and 89 and 75 is halfway between 70 and 79.

Therefore, we can also condense the 25 individual grades into five groups each of which is represented by a number grade. Thus, we can also say that in this English class there are five 95's, six 85's, five 75's, six 65's, and three 55's.

This condensation of our data from 25 numbers to 5 numbers will cause us to lose accuracy. However, we are willing to sacrifice some accuracy in order to simplify our mathematics.

For example, if we added all of the 25 individual grades together and then divided the sum by 25, we would get:

$$\overline{X} = \frac{\Sigma X}{N} = \frac{1900}{25} = 76$$

Data grouping will give us:

$$\overline{X} = \frac{\Sigma FX}{N} = \frac{(5 \times 95) + (6 \times 85) + (5 \times 75) + (6 \times 65) + (3 \times 55)}{5 + 6 + 5 + 6 + 3}$$

$$= \frac{475 + 510 + 375 + 390 + 165}{25} = \frac{1915}{25} = 76.6$$

For most practical purposes, 76.6 is close enough to 76 to make data grouping worthwhile. We are actually interested in trends and not in exact values. Our original data is probably not that exact to begin with.

In statistics, the groups are called class intervals, and the group boundaries are called class boundaries. The class midpoint is found by adding the lower class boundary of the class interval to the lower class boundary of the next largest class interval, and then dividing this sum by 2. Thus, in our problem above, the class midpoint of the class interval 80—89 is equal to:

$$\frac{80 + 90}{2} = 85$$

1. <u>PROCEDURE FOR CONDENSING DATA INTO CLASS INTERVAL: First</u>, find the largest and smallest numbers in the data, and then subtract them in order to find the range. <u>Secondly</u>, divide the range into a convenient number of groups — usually 5, 10, 15 or 20. Ten is probably the most common number of groups or class intervals used. <u>Next</u>, we determine the class boundaries, taking care to avoid any overlapping. A number from the data should be able to go into only one of the class intervals.

For example, if we had the class intervals, 30 - 60, 60 - 90, 90 - 120, the number 60 could go into either one of two intervals. This is not allowed, so we can change these intervals as follows: 30 - 59.9, 60 - 89.9, 90 - 119.9. <u>Finally, all class intervals must be the same size.</u>

2. <u>SAMPLE CLASS INTERVAL PROBLEMS</u>.

EXAMPLE 1: Arrange the following numbers into five class intervals: 0, 1, 9, 6, 3, 1, 8, 0, 1, 3, 7, 0, 6, 5, 4, 7, 6, 2, 9, 2, 2, 4, 2, 3,3, 5, 3, 2, 7, 1, 8, 1, 7, 6, 5, 7, 6, 1, 0 and 9:

Solution

Step 1. Range = (X max, –Xmin.) = (9–0) = 9

Step 2. Size of class interval

$$= \frac{\text{Range}}{\text{Number of intervals desired}}$$

$$= \frac{9}{5} = 1.8; \text{ call it } 2$$

We rounded off to two because we want an easy number to work with.

Step 3. The lowest number in the data is zero. This will therefore be the lower boundary of the first class interval. The size of the class interval is 2. This means that there will be two numbers in each interval. In the first interval therefore, we will have the numbers 0 and 1. The second interval consequently must contain the numbers 2 and 3. The third interval will have 4 and 5. Thus, the lower boundaries of the class intervals are:

0 - 1.9	2 - 3.9	4 - 5.9	6 - 7.9	8 - 9.9

Step 4. Arrange the data into the class intervals:

Intervals	Individual Numbers in the Class Interval	Number of Items in the Class Interval (Frequency)	Midpoint of Class Interval
0 - 1.9	0, 1, 1, 0, 1, 0, 1, 1, 1, 0	10	1
2 - 3.9	3, 3, 2, 2, 2, 2, 3, 3, 3, 2	10	3
4 - 5.9	5, 4, 4, 5, 5	5	5
6 - 7.9	6, 7, 6, 7, 6, 7, 7, 6, 7, 6	10	7
8 - 9.9	9, 8, 9,8, 9	5	9

Step 5. Count the number of items in Column 2 to get the frequency for each class interval listed in Column 3.

Step 6. The class midpoints in Column 4 were obtained by adding the lower class boundary of the next largest class interval:

Class Interval	Calculation	Midpoint
0 - 1.9	$\frac{0+2}{2}=\frac{2}{2}$	= 1
2 - 3.9	$\frac{2+4}{2}=\frac{6}{2}$	= 3
4 - 5.9	$\frac{4+6}{2}=\frac{10}{2}$	= 5
6 - 7.9	$\frac{6+8}{2}=\frac{14}{2}$	= 7
0 - 9.9	$\frac{8+10}{2}=\frac{18}{2}$	= 9

EXAMPLE 2: Arrange the following numbers into ten class intervals:

95	215	100	205	65	80	110	205
165	125	225	125	110	190	185	170
115	130	140	120	115	125	110	185
120	80	80	130	175	85	105	110
180	80	135	125	90	115	100	167

Solution

Step 1. Range = (X max. – X min.) = (225 – 65) = 160

Step 2. Size of class interval

$$= \frac{\text{Range}}{\text{number of intervals desired}}$$

$$= \frac{160}{10} = 16$$

Round off to 20 to make it easier to determine the class intervals.

Step 3. The lowest number in the data is 65, so this will be the lower boundary of the first class interval. The size of the class interval is 20, so the next lower boundary is (65 + 20) = 85. Thus, the lower boundaries of the class intervals are:

65 -
85 -
105 -
125 -
145 -
165 -
185 -
205 -
225 -
245 -

The upper boundary for a class interval should be a little less than the lower boundary for the next class interval in order to avoid overlapping. Therefore, our ten class intervals are:

65 - 84.9
85 - 104.9
105 - 124.9
125 - 144.9
145 - 164.9
165 - 184.9
185 - 204.9
205 - 224.9
225 - 244.9
245 - 264.9

Step 4. Arrange the data into the class intervals:

Class Intervals	Number of Items in the Class Interval (Frequency)		Midpoint of the Class Interval
65 - 84.9	𝍸	5	75
85 - 104.9	𝍸	5	95
105 - 124. 9	𝍸 𝍸	10	115
125 - 144. 9	𝍸 111	8	135
145 - 164.9		0	155
165 - 184.9	𝍸	5	175
185 - 204. 9	111	3	195
205 - 224. 9	111	3	215
225 - 244. 9	1	1	235
245 - 264. 9		0	255

Instead of inserting the individual numbers in Column 2, we used a tally sheet. That is, as we read through the numbers in the data, we made a mark in Column 2 for each number that went into a class interval. Then we counted up the number of marks to get the number of items in the interval. This step combines Steps 4 and 5 in the previous example.

Step 5. Calculation of class midpoints:

Class Intervals	Calculation	Midpoint
65 - 84. 9	$\frac{65+85}{2}=\frac{150}{2}=$	75
85 - 104.9	$\frac{85+105}{2}=\frac{190}{2}=$	95
105 - 124.9	$\frac{105+125}{2}=\frac{230}{2}=$	115
125 - 144. 9	$\frac{125+145}{2}=\frac{270}{2}=$	135
145 - 164. 9	$\frac{145+165}{2}=\frac{310}{2}=$	155
165 - 184. 9	$\frac{165+185}{2}=\frac{350}{2}=$	175
185 - 204.9	$\frac{185+205}{2}=\frac{390}{2}=$	195
205 - 224. 9	$\frac{205+225}{2}=\frac{430}{2}=$	215
225 - 244. 9	$\frac{225+245}{2}=\frac{470}{2}=$	235
245 - 264. 9	$\frac{245+265}{2}=\frac{510}{2}=$	255

COMPUTING THE MEAN FROM GROUPED DATA. If we multiply and then divide an equation by the same number, it would have no effect on the equation, since a number divided by itself is always equal to 1. For example:

$$\frac{9}{9}=1,\ \frac{76}{76}=1,\ \text{etc.}$$

In the equation Y = 4X, if we multiplied the 4X by 6 and then divided the result by 6, we would end up with 4X:

$$6\,(4X) = 24X,\ \frac{24X}{6} = 4X$$

In the section on how to use an assumed average to compute the mean from ungrouped data, we developed the formula:

$$\overline{X} = A + \frac{\Sigma\ (X - A)}{N}$$

If, for example, we multiplied the term:

$$\frac{\Sigma\ (X - A)}{N} \text{ by i}$$

and then divided it by i, we would get:

$$\frac{\Sigma\ (X - A)}{N} \cdot \frac{i}{i}$$

We can rearrange this term to get:

$$\frac{\Sigma\ (X - A)}{\frac{i}{N}}\ i$$

If we canceled out the i's we would return to our original term:

$$\frac{\Sigma\ (X - A)}{N}$$

We let the letter i represent the size of the class interval. The value for i is found by subtracting the lower class boundary of the class interval from the lower class boundary of the next largest class interval. We will have to find the value for i only once in a problem, since all class intervals should be the same size. Thus, if we had the class intervals: 30 - 59.9, 60 -89.9, 90 -119.9,120-149.9, i would be equal to (60–30) = 30.

Actually, i is the range of the class interval. The size of the class interval i is also equal to the difference between two adjacent midpoints. Thus, for the intervals 30 - 59.9, 60 - 89.9, and 90 - 119.9, the midpoints are 45, 75, and 105 respectively. Therefore, i = (75 – 45) = 30

Now if we let $d = \frac{X - A}{i}$, our formula for the mean becomes equal to;

$$\overline{X} = A + \frac{\Sigma d}{N}\ i \text{ and}$$

$$\overline{X} = A + \frac{\Sigma Fd}{N}\ i$$

In grouped data, the X in the (X – A) term is equal to the class midpoint. Remember that all the numbers in the class interval are now equal to the class midpoint. The A, in the X – A term is equal to the class midpoint which was picked as the assumed average.

The term $\frac{X - A}{i}$ tells us how many times i goes into X – A. In other words, it tells us how many class intervals away an X – A term is from the assumed average. So d, since it equals $\frac{X - A}{i}$, represents the number of i's away that a class interval is from the assumed average.

The procedure for using the formula is:

$$\overline{X} = A + \frac{\Sigma Fd}{N}\ i$$

called the short cut method of calculating the mean. It is illustrated in the following example.

EXAMPLE: Arrange the following data into 10 class intervals and compute the mean.

2, 15, 22, 76, 54, 98, 53, 22, 68, 0, 5, 5, 31,45, 48, 42, 12, 26, 37, 6, 22, 34, 12, 18 and 10:

Solution

Step 1. Range = (98 – 0) = 98

Step 2. Size of class interval = $\frac{\text{range}}{\text{no. of intervals}}$

$$i = \frac{98}{10} = 9.8$$

Let i = 10 to make it easier to work with.

Step 3. Lowest number = 0, so this will be the first lower class boundary. The class size = 10, so lower class boundaries will be 0 -, 10 -, 20 -, 30 -, 40 -, 50 -, 60 -, 70 -, 80 -, 90 -

The upper boundary is a little less than the lower boundary for the next class interval. Therefore, our 10 class intervals are: 0 to 9.9, 10 to 19.9, 20 to 29.9, 30 to 39.9, 40 to 49.9, 50 to 59.9, 60 to 69.9, 70 to 79.9, 80 to 89.9, and 90 to 99.9.

Step 4. Construct a table and arrange the data into the class intervals:

Class Interval	Tally	Frequency F	Midpoint Xm
0 - 9.9	~~1111~~	5	5
10- 19.9	~~1111~~	5	15
20 - 29.9	1111	4	25
30 - 39.9	111	3	35
40 - 49.9	111	3	45
50 - 59.9	11	2	55
60 - 69.9	1	1	65
70 - 79.9	1	1	75
80 - 89.9		0	85
90 - 99.9	1	1	95

Step 5. Assume A = 45. Remember, we must pick one of the class midpoints as an assumed average. Any midpoint could have been chosen. We picked 45 because it is close to the middle class interval and because it will help us illustrate the shortcut method of calculating the mean:

Class Interval	Frequency	Midpoint	Assumed Average	Midpoint Minus Assumed Average	$\frac{Xm - A}{i} = d$	Frequency Times d
	F	Xm	A	(Xm – A)	d	Fd
Col 1	Col 2	Col 3	Col 4	Col 5	Col 6	Col 7
0 - 9.9	5	5		−40	−4	−20
10 - 19.9	5	15		−30	−3	−15
20 - 29. 9	4	25		−20	−2	−8
30 - 39. 9	3	35		−10	−1	−3
40 - 49. 9	3	45	45	0	0	0
50 - 59. 9	2	55		+10	+1	+2
60 - 69.9	1	65		+20	+2	+2
70 - 79. 9	1	75		+30	+3	+3
80 - 89. 9	0	85		+40	+4	+0
90 - 99. 9	1	95		+50	+5	+5

Step 6. Column 5 = (Column 3 – Column 4)
(Xm – A) = Xm – A = Xm – 45
Column 6 = Column 5 ÷ i
d = (Xm – A) ÷ i
Column 7 = Column 2 × Column 6
Fd = FXd

NOTE: Always multiply across the table.

Step 7. $\overline{X} = A + \frac{\Sigma Fd}{N} i$

A = 45

ΣFd = sum of Fd Column 7 = – 20 – 15 –8 –3 +0 + 2 + 2 + 3 + 0 + 5 = -34
(Watch your signs)

i = 10

N = ΣF = sum of F column = = 5 + 5 + 4 + 3 + 3 + 2 + 1

(We started out with 25 numbers.)

$$\overline{X} = \frac{-34}{25}(10) = 45 + \frac{-340}{25}$$

$$\overline{X} = 45 - 13.6 = 31.4$$

Column 6, which was equal to $\frac{(Xm - A)}{i} = d$ also represents the number of class intervals that a given class interval is away from the assumed average. For example, the –4 means that the first class interval is 4 intervals from the assumed average. The minus sign indicates that its values are smaller than the assumed average. You can see that Column 6 counts up, from the assumed average, 1, 2, 3, 4, 5 intervals and –1, –2, –3, –4 intervals down from the assumed average. Thus, d is actually the distance from the assumed average measured in class intervals.

We can, therefore, omit the need for finding Xm and (Xm–A) by just counting the class intervals from the assumed average to get the d's.

For example, let us do the above problem again, taking 25 as the assumed average. For the d Column we let the row containing A. Then we count the class intervals up and down from A or 0. Minus means less than A and plus means greater than A:

Class Interval	**Frequency**	**Assumed Average**	**d**	**Fd**
Column 1	Column 2	Column 3	Column 4	Column 5
0 - 9.9	5		−2	−10
10 - 19.9	5		−1	−5
20 - 29. 9	4	25	0	0
30 - 39. 9	3		+1	3
40 - 49. 9	3		+2	6
50 - 59. 9	2		+3	6
60 - 69.9	1		+4	4
70 - 79. 9	1		+5	5
80 - 89. 9	0		+6	0
90 - 99. 9	1		+7	7

Column 5 = Column 2 times Column 4

$$\overline{X} = A + \frac{\Sigma Fd}{N} i$$

A = 25

i = 10

ΣFd = sum of F × d column = −10 – 5 + 0 + 3 + 6 + 6 + 4 + 5 + 0 + 7 = +16
(Watch your signs)

N = ΣF = sum of F column

= 5 + 5 + 4 + 3 + 3 + 2 + 1 + 1+ 0 + 1

= 25

$$\overline{X} = A + \frac{\Sigma Fd}{N} i$$

$$\overline{X} = 25 + \frac{16}{25}(10) = 25 + \frac{160}{25} = 25 + 6.4$$

$$\overline{X} = 31.4$$

Now let us do the same problem taking 65 as the assumed average:

Class Interval	Frequency	Assumed Average	d	Fd
Column 1	Column 2	Column 3	Column 4	Column 5
0 - 9.9	5		−6	−30
10 - 19.9	5		−5	−25
20 - 29. 9	4		−4	−16
30 - 39. 9	3		−3	−9
40 - 49. 9	3		−2	−6
50 - 59. 9	2		−1	−2
60 - 69.9	1	65	0	0
70 - 79.9	1		+1	1
80 - 89. 9	0		+2	0
90 - 99.9	1		+3	3

$$\overline{X} = A + \frac{\Sigma Fd}{N} i$$

A = 65 to get d we count the intervals from A, where
d = 0 at A.

ΣFd = sum of Fd column = $(-30 -25 -16 - 9 - 6 - 2 + 0 + 1 + 0 + 3 = -84$

(Watch your signs)

i = 10

N = ΣF = sum of F column = 25

$$\overline{X} = 65 - \frac{84}{25}(10) = 65 - \frac{840}{25}$$

$$\overline{X} = 65 - 33.6 = 31.4$$

The answer always comes out the same, no matter which midpoint was picked as the assumed average. Remember, to get d we just count the number of intervals away that the class interval is from assumed average. Minus means that the interval is smaller than A and vice versa.

COMPUTING THE STANDARD DEVIATION FROM GROUPED DATA. In the previous section, we developed a formula and a procedure for calculating the mean from grouped data. This method can be extended to the standard deviation. The formula for computing the standard deviation from grouped data is:

$$\sigma = \sqrt{\left(\frac{\Sigma Fd^2}{N} i^2\right) - \left(\frac{\Sigma Fd}{N} i\right)^2}$$

and the table headings are:

Class Interval	Frequency	Assumed Average	d	Fd	Fd^2
Col. 1	Col. 2	Col. 3	Col. 4	Col. 5	Col. 6

Column 5 = Column 2 × Column 4

$$Fd = F \times d$$

Column 6 = Column 5 × Column 4

$$Fd^2 = Fd \times d$$

For 5 we just add one additional Column, Fd_2 to the $\overline{X}$ table

Example: Find the standard deviation for the example in the previous section, which illustrates the computation of the mean from grouped data. The table is given below:

Class Interval	Frequency	Assumed Average	d	Fd	Fd^2
Col. 1	Col. 2	Col. 3	Col. 4	Col. 5	Col. 6
0 - 9.9	5		−6	−30	+180
10 - 19.9	5		−5	−25	+125
20 - 29. 9	4		−4	−16	+64
30 - 39. 9	3		−3	−9	+27
40 - 49. 9	3		−2	−6	+12
50 - 59.9	2		−1	−2	2
60 - 69. 9	1	65	0	0	0
70 - 79. 9	1		+1	1	1
80 - 89. 9	0		+2	0	0
90 - 99. 9	1		+3	3	9

Step 1. Column 5 = Column 2 × Column 4

Fd = F times d

$\Sigma Fd = -30 -25 -16 -9 -6 -2 + 0 + 1 + 0 + 3$

$\Sigma Fd = -84$

(Multiply Across)

Step 2. Column 6 = Column 4 × Column 5

$Fd^2 = Fd \times d$

$\Sigma Fd^2 = (180 + 125 + 64 + 27 + 12 +2 + 0 + 1 + 0 + 9)$

$\Sigma Fd^2 = 420$

(Multiply across the table)

Step 3. N = ΣF = sum of Column 2

N = (5 + 5 + 4 + 3 + 3 + 2 + 1 + 1 + 0 + 1)

N = 25

Step 4. i = range of the class interval

i = (10 − 0) = 10

Step 5. $\sigma = \sqrt{\left(\frac{\Sigma Fd^2}{N} i^2\right) - \left(\frac{\Sigma Fd}{N} i\right)^2}$

$$\sigma = \sqrt{\left(\frac{420(100)}{25}\right) - \left(\frac{-84(10)}{25}\right)^2}$$

$$\sigma = \sqrt{\left(\frac{42000}{25}\right) - \left(\frac{-840}{25}\right)^2}$$

$$\sigma = \sqrt{(1680) - (-33.6)^2}$$

$$\sigma = \sqrt{(1680) - (1128.95)}$$

$$\sigma = \sqrt{551.04}$$

$$\sigma = 23.5$$

PROBLEM 1: Compute the mean and standard deviation for the following data on the weight in grams of 50 oranges:

100, 107, 180, 188, 218, 210, 205, 109, 120, 110, 270, 600, 265, 260,127, 128, 129, 130, 130, 130, 128, 142, 149, 143, 152, 153, 154, 154, 155, 155, 110, 170, 171, 263, 170, 228, 229, 230, 232, 236, 242, 244, 246, 234, 235, 236, 238, 240, 242, and 236

Solution

Step 1. The lowest number in the data is 100 and the highest is 600. If we examine the data, we can see that the number 600 is out of place. It is too extreme. If we used this number to compute class intervals, thent the intervals from 270 to 600 would have no numbers in them. The next highest number is 270. It is more representative of the data than 600. We can use 270 to compute class intervals and make the highest interval open ended to include the number 600.

Range for computing the class interval = (100 – 270) = 170

Step 2. Compute the size of the class interval. Let's use 7 intervals to make the problem easier to work with.

$$i = \frac{170}{7} = 24.3. \text{ Let } i = 30$$

Step 3. Determine the class intervals.

Lowest number = 100, so we can let this be our lower class boundary. The class size i = 30, so the lower class boundaries will be: 100 -, 130 -, 160 -, 100 -, 220 -, 250 -, and 280 -

The upper boundary is a little less than the lower boundary for the next highest class interval. Therefore, the 7 class intervals are: 100 - 129. 9, 130 - 159.9, 160 - 189. 9, 190 - 219.9, 220 - 249.9, 250 - 279.9 and over 280

The last class interval is "open ended" to include all numbers over 280

Step 4. Construct a table and arrange the data into class intervals:

Class Interval	Numbers in the Interval	Fre-quency	Assumed Average	d	Fd	Fd^2
Col. 1	Col. 2	Col. 3	Col. 4	Col. 5	Col. 6	Col. 7
100 -129.9	100, 107, 109, 110, 127, 128, 129, 128, 110, 120.	10		–4	–40	+160
130 -159.9	130, 130, 130, 142, 149, 143, 152, 153, 154, 154, 155, 155.	12		–3	–36	+108
160 -189.9	180, 188, 170, 171, 170.	5		–2	–10	+20
190 -219.9	218, 210, 205.	3		–1	–3	+3
220 -249.9	228, 229, 230, 232, 242, 244, 246, 234, 235, 236, 236, 236, 238, 240, 242.	15	235	0	0	0
250 -279.9	270, 265, 260, 263.	4		1	4	4
Over 280	600	1		2	2	4
	TOTALS	50			–83	299

Column 6 = Column 3 times Column 5

Column 7 = Column 5 times Column 6

Remember to get d, we just count the class intervals up and down from A, At Ad = 0, minus d means that the interval is smaller than the A interval and vice versa.

Step 5. Compute $\overline{X}$

$$\overline{X} = A + \frac{\Sigma Fd}{N} i$$

N = ΣF = sum of Column 3 = 50

A = assumed average = 235

(We picked the midpoint of the class interval having the highest frequency as the assumed average. Any midpoint could have been picked as A).

i = range of class intervals = 30

$$\overline{X} = 235 + \frac{-83}{50}(30)$$

$$\overline{X} = 235 - 49.8 = 185.2 \text{ grams}$$

Step 6. Compute σ

$$\sigma = \sqrt{\left(\frac{\Sigma Fd^2}{N} i^2\right) - \left(\frac{\Sigma Fd}{N} i\right)^2}$$

ΣFd^2 = sum of Column 7

= 299

i^2 = (i) (i) = (30) . (30) = 900

N = ΣF= sum of Column 3 = 50

ΣFd sum of Column 6 = – 83

i = 30

$$\sigma = \sqrt{\left(\frac{299}{50}(900)\right) - \left(\frac{-83 \times 30}{50}\right)^2}$$

$$\sigma = \sqrt{(5382) - (-49.8)^2}$$

$$\sigma = \sqrt{(5382 - 2480)} = \sqrt{2902}$$

σ =53.9 grams

PROBLEM 2: The following data represents the amount of time, in seconds, that it took 25 high school students to run the 100 yard dash. Arrange times into 5 groups. The first group should contain the times under 10 seconds and the last group should contain the times over 25 seconds. Next compute the average time and the standard deviation:

9.7, 11.2, 11.3, 12.1, 22.5, 18.1, 18.1, 18.5,18.2, 16.0, 16.2, 16.7, 17.5, 17.6, 17.8, 12.5, 12.6, 12.8, 13.0, 13.5, 13.2,13.7, 28.0, 26.1, and 40.5

Solution

Step 1. The problem requires 5 groups or class intervals. The first interval contains the times under 10 seconds. The fifth interval contains the times over 25 seconds. There are 3 intervals between 10 and 25. Therefore, there are (25 – 10) = 15 seconds to be spread between 3 intervals.

Step 2. Compute the size of the class interval:

$$i = \frac{15 \text{ seconds}}{3 \text{ intervals}} = \frac{5 \text{ seconds}}{1 \text{ interval}}$$

Step 3. The class intervals are:

under 10 seconds
10 - 14.9
15 - 19.9
20 - 24.9
over 25 seconds

Step 4. Construct a table and arrange the data into the class intervals:

Class Interval	Tally	Frequency	Assumed Average	d	Fd	Fd^2
Col. 1	Col. 2	Col. 3	Col. 4	Col. 5	Col. 6	Col. 7
Under 10	1	1		−2	−2	4
10 - 14.9	~~1111~~ ~~1111~~	10		−1	−10	10
15 - 19.9	~~1111~~ 1111	9	17.5	0	0	0
20 - 24.9	11	2		1	2	2
Over 25	111	3		2	6	12
TOTALS		25			−4	28

Column 6 = (Column 3 times Column 5)

Column 7 = (Column 6 times Column 5)

Step 5. Compute $\overline{X}$:

$$\overline{X} = A + \frac{\Sigma Fd}{N} i$$

N = ΣF = sum of Column 3 = 25

ΣFd = sum of Column 6 = −4

A = assumed average = 17.5

(We picked the class interval having F = 9 instead of F = 10 because it is easier to multiply the d terms by 10 than by 9. Picking A as the midpoint of the F = 9 interval eliminates this frequency from the calculations.)

i = range of the class interval = 5

$$\overline{X} = 17.5 + \frac{-4}{25}(5)$$

$$\overline{X} = (17.5 - .8) = 16.7 \text{ seconds}$$

Step 6. Compute $\sigma = \sqrt{\left(\frac{\Sigma Fd^2}{N} i^2\right) - \left(\frac{\Sigma Fd}{N} i\right)^2}$

ΣFd^2 = sum of column 7 =28

i^2 = (i) (i) =(5) (5) =25, N =25

ΣFd = −4 x 5 = -20

$$\sigma = \sqrt{\left(\frac{28}{25}(25)\right) - \left(\frac{-4 \times 5}{25}\right)^2}$$

$$\sigma = \sqrt{28 - (-.8)^2}$$

$$\sigma = \sqrt{28 - .64} = \sqrt{27.36}$$

$$\sigma = 5.2 \text{ seconds}$$

NOTE: In the grouped data calculations, the standard deviation can be found without the need for computing $\overline{X}$ first.

6 FREQUENCY DISTRIBUTION

HOW TO DRAW VARIOUS GRAPHS OF FREQUENCY DISTRIBUTIONS

INTRODUCTION TO FREQUENCY HISTOGRAMS. A frequency histogram is simply a bar chart of class frequency versus class interval. It is a pictorial method of showing how the data is distributed among the class intervals.

1. CLASS INTERVALS. The procedure for condensing data into class intervals is covered in Topic #5.

2. PROCEDURE FOR CONSTRUCTING A FREQUENCY HISTOGRAM. In constructing a frequency histogram, the vertical axis represents the class frequency and the horizontal axis represents the class intervals. The class frequencies are represented by vertical bars. Each bar is one class interval wide, and is centered over its class midpoint. The height of the bar is equal to the class frequency of its corresponding class interval. The vertical axis always starts with zero to illustrate the relative size of the bars.

   The procedure for constructing a frequency histogram will be illustrated by the solution of a Sample problem.

EXAMPLE: Draw a frequency histogram for the following data:

Class Interval	Frequency
5 - 10 lbs.	2
10 - 15 lbs.	8
15 - 20 lbs.	4
20 - 25 lbs.	3

Solution A frequency histogram is a graph of frequency versus class interval. Let the vertical axis = frequency and the horizontal axis = class interval (weight in pounds).

Step 1. Divide the horizontal scale into equal lengths, one for each class interval. There are four class intervals in this example:

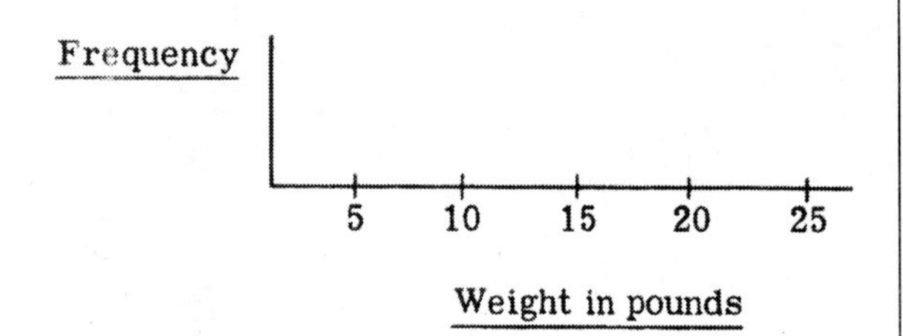

Step 2. Divide the vertical scale into equal lengths to represent all the frequencies starting with zero and going to the maximum value in the data:

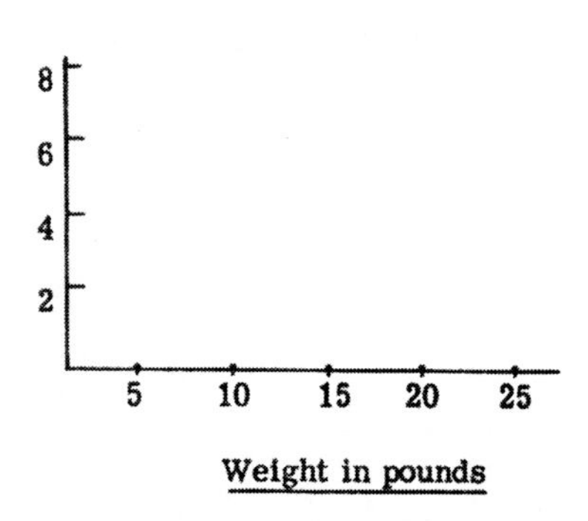

Step 3. Plot bar graphs of the data to show how the various weights are distributed:

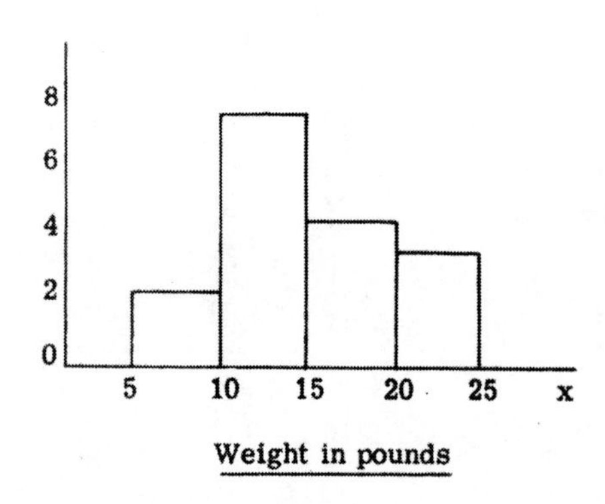

3. SAMPLE FREQUENCY HISTOGRAM PROBLEMS.

PROBLEM 1: Draw a frequency histogram for the following data:

Class Interval	Frequency
.002 - .004 inches	5
.004 - .006 inches	5
.006 – .008 inches	10
.008 – .010 inches	4
.010 – .012 inches	7

Solution

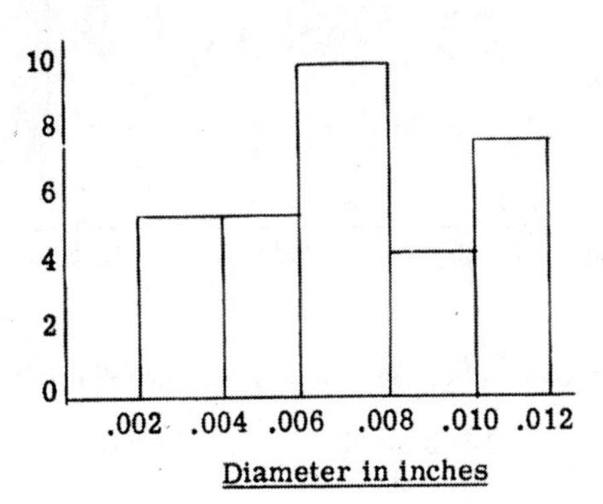

PROBLEM 2: Draw a frequency histogram for the following data:

Class Interval	Frequency
99 - 100	3
101 - 102	7
103 - 104	9
105 - 106	12

Solution

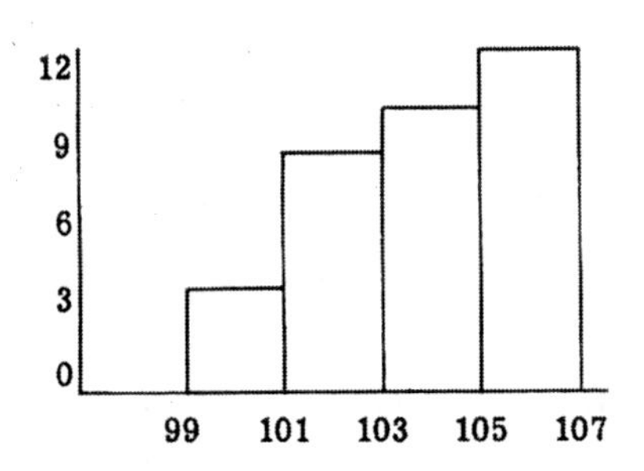

INTRODUCTION TO THE FREQUENCY POLYGON. A frequency polygon is a "broken line" graph of class frequency versus midpoint of the class interval.

1. PROCEDURE FOR CONSTRUCTION OF A FREQUENCY POLYGON. The procedure for constructing a frequency polygon is similar to that of the histogram. The vertical axis represents the class frequency and the horizontal axis represents the values of the midpoints of the class intervals. In the frequency polygon, the number of class intervals is increased by adding two class intervals, each having zero frequency. One interval is added just below the lowest interval and the other interval is added just above the largest interval. This is to allow the frequency polygon to start and end at a zero frequency point.

The plotted frequencies are connected by a series of straight lines which go from one adjacent point to the other, forming a many-sided figure. The word, polygon, means many-sided.

The procedure for constructing a frequency polygon will be illustrated by the solution of a sample problem.

EXAMPLE: Draw a frequency polygon for the following data:

Class Interval	Class Midpoint	Frequency
9 - 11.9	10.5	1
12 - 14.9	13.5	2
15 - 17.9	16.5	5
18 - 20.9	19.5	3

Solution A frequency polygon is a graph of frequency versus class midpoint. In order to start and end the graph at zero, two class intervals, each having zero frequency, have to be added to the beginning and end of the data.

Step 1. Add class intervals 6 - 8.9 and 21 - 23.9, with zero frequencies, to the data:

Class Interval	Class Midpoint	Frequency
6 - 8.9	7.5	0
9 - 11.9	10.5	1
12 - 14.9	13.5	2
15 - 17.9	16.5	5
18 - 20.9	19.5	3
21 - 23.9	22.5	0

Step 2. Let the vertical axis = frequency and the horizontal axis = class midpoint:

Divide the horizontal scale into equal lengths one for each class midpoint. There are six class midpoints in this example:

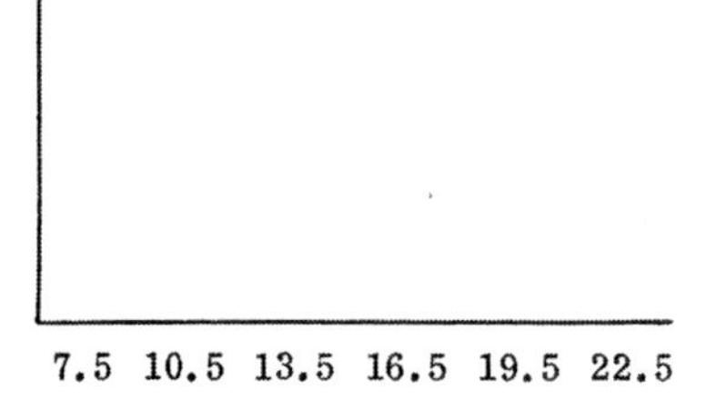

Step 3. Divide the vertical scale into equal lengths to represent all the frequencies, starting with zero and going on to the maximum value in one data:

Frequency

5
4
3
2
1
0

7.5 10.5 13.5 16.5 19.5

Step 4. Plot the frequency points on the graphs, and then connect the adjacent points with a straight line:

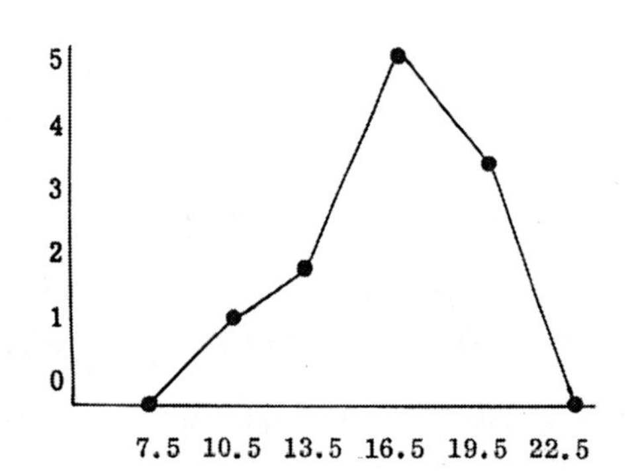

2. SAMPLE FREQUENCY POLYGON PROBLEMS.

PROBLEM 1: Draw a frequency polygon for the following data:

Class Interval	Frequency
.002 - .003	5
.004 - .005	5
.006 - .007	10
.008 - .009	4
.010 - .011	7

Solution

Class Interval	Class Midpoint	Frequency
.000 - .0019	.0005	0
.002 - .0039	.0025	5
.004 - .0059	.0045	5
.006 - .0079	.0065	10
.008 - .0099	.0085	4
.010 - .0119	.0105	7
.012 - .0139	.0125	0

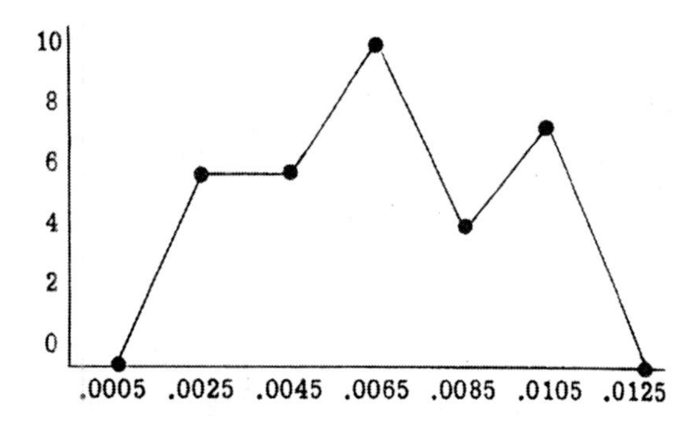

PROBLEM 2: Draw a frequency polygon for the following data:

Class Interval	Frequency
100 - 200	25
200 - 300	50
300 - 400	40

Solution

Class Interval	Class Midpoint	Frequency
0 - 100	50	0
100 - 200	150	25
200 – 300	250	50
300 – 400	350	40
400 - 500	450	0

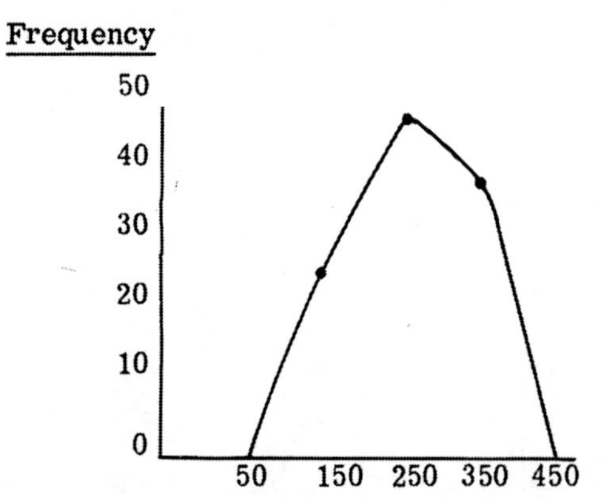

INTRODUCTION TO THE FREQUENCY CURVE. A frequency curve is a frequency polygon with a smooth continuous curve through the plotted points of the graph of class frequency versus class midpoint.

1. PROCEDURE FOR CONSTRUCTING A FREQUENCY CURVE. The procedure for constructing a frequency curve is the same as the procedure for constructing a frequency polygon. The difference is that plotted frequencies are connected by a smooth continuous curve instead of by a series of straight lines.

2. SAMPLE FREQUENCY CURVE PROBLEMS.

PROBLEM 1: Draw a frequency curve for the following data:

Class Interval	Frequency
5 - 9	2
10 - 14	8
15 - 19	4

20 - 24	3

Solution

Class Interval	Class Midpoint	Frequency
0 - 4.9	2.5	0
5 - 9.9	7.5	2
10 - 14.9	12.5	8
15 - 19.9	17.5	4
20 - 24.9	22.5	3
25 - 29.9	27.5	0

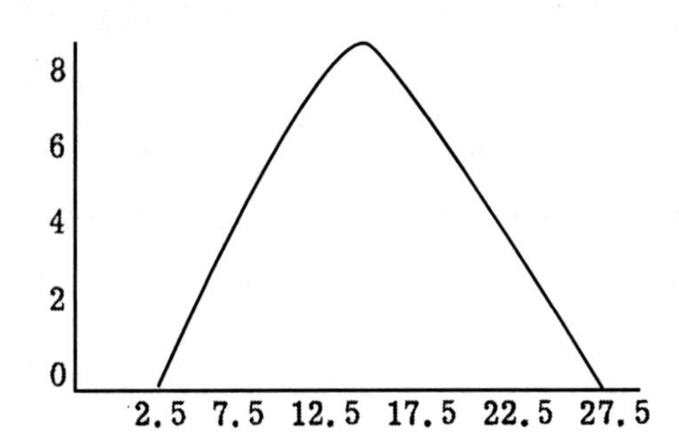

PROBLEM 2: Draw a frequency curve for the following data:

Class Interval	Frequency
99 - 100	3
101 - 102	7
103 - 104	9
105 - 106	12

Solution

Class Interval	Class Midpoint	Frequency
97 - 98	98	0
99 - 100	100	2
101 - 102	102	7
103 - 104	104	9
105 - 106	106	12
107 - 108	108	0

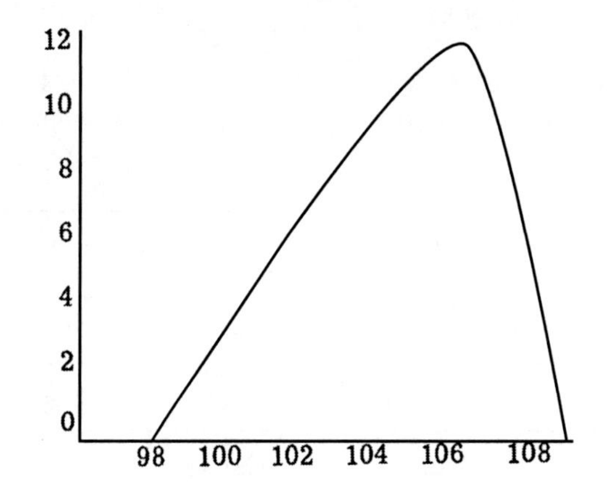

7 PERMUTATIONS AND COMBINATIONS

INTRODUCTION TO PERMUTATIONS. A permutation is defined as the number of different possible ways that we can arrange a group of items in a definite order.

For example, in how many different ways can a History book, an English book and a Statistics book be arranged in three empty spaces on a shelf?

Arrangement Number	Book in First Space	Book in Second Space	Book in Third Space
1	Statistics	English	History
2	Statistics	History	English
3	English	History	Statistics
4	English	Statistics	History
5	History	Statistics	English
6	History	English	Statistics

We can see that there are 6 possible ways in which the 3 books can be arranged.

1. BOX METHOD OF CALCULATING PERMUTATIONS.
   The permutations of any number of items can be found by simply writing out all the possible arrangements and then counting them. This procedure can be very time-consuming, particularly if we are going to determine the number of arrangements for a large group of items. There is a simpler technique for finding the number of permutations. It is called the "box method." This method can best be explained through the solution of a series of typical permutation problems.

2. SAMPLE "BOX METHOD" PERMUTATION PROBLEMS.

EXAMPLE 1: In our book example, the problem was to fill three spaces on a shelf. These spaces can be represented by three boxes, $\boxed{\ }\boxed{\ }\boxed{\ }$. Let us look at this problem from the point of view of the spaces on the shelf. Space #1 has a choice of three books: Statistics, English, or History. So we put a 3 in the first box $\boxed{3}\boxed{\ }\boxed{\ }$. After space #1 has been picked, there are only two books left. Space #2 now picks. It has a choice of only two books to pick from so we put a 2 in the second box $\boxed{3}\boxed{2}\boxed{\ }$. Space #1 picked and space #2 picked, so now there is only one book left. Space #3 picks. It has a choice of only one book, the book that is left, so we put a 1 in the third box $\boxed{3}\boxed{2}\boxed{1}$. If we multiply these three numbers together, we get $3 \times 2 \times 1 = 6$. This is equal to the number of permutations for the three books.

EXAMPLE 2: Suppose we had five different books and wanted to pick out three of them to arrange in three spaces on a shelf. How many different arrangements or permutations of the three books can we get? Again let's look at this problem from the point of view of the spaces on the shelf. The three spaces can be represented by 3 boxes, $\boxed{\ }\boxed{\ }\boxed{\ }$. Space #1 has a choice of any of the 5 books. So we put a 5 in the first box $\boxed{5}\boxed{\ }\boxed{\ }$. After space #1 picked, there are 4 books left to pick from. Space #2 now picks. It has a choice of 4 books so we put a 4 in the second box $\boxed{5}\boxed{4}\boxed{\ }$. Now there are only 3 books left. Space #3 picks. It has a choice of any of the three remaining books so we put a 3 in the third box $\boxed{5}\boxed{4}\boxed{3}$. The total number of permutations that we can get of 5 books taken three at a time is thus $5 \times 4 \times 3 = 60$

EXAMPLE 3: In how many ways can ten people line up at a ticket office? The ten places on line can be represented by ten boxes $\boxed{\ }\boxed{\ }\boxed{\ }\boxed{\ }\boxed{\ }\boxed{\ }\boxed{\ }\boxed{\ }\boxed{\ }\boxed{\ }$. The first place on line has a choice of any one of the ten people, so we put a 10 in the first box $\boxed{10}\boxed{\ }\boxed{\ }\boxed{\ }\boxed{\ }\boxed{\ }\boxed{\ }\boxed{\ }\boxed{\ }\boxed{\ }$. The first space has been filled, so now there are 9 people left. Any one of the 9 remaining people can go in the second place so we put a 9 in the second box $\boxed{10}\boxed{9}\boxed{\ }\boxed{\ }\boxed{\ }\boxed{\ }\boxed{\ }\boxed{\ }\boxed{\ }\boxed{\ }$. We can see that if we continue this reasoning for the remaining people we will end up with the 10 boxes filled as follows:

|10|9|8|7|6|5|4|3|2|1| .

Therefore, the total number of permutations that we can get of ten things taken 10 at a time is:

$10 \times 9 \times 8 \times 7 \times 6 \times 5 \times 4 \times 3 \times 2 \times 1 = 3{,}628{,}800$

EXAMPLE 4: A newsstand sells 6 different papers. In how many different ways can these papers be distributed among 4 people, if each person gets a different paper ? In this example we have to associate the 6 papers with the 4 people. Therefore, the boxes will represent the people. There are 4 boxes | | | | |. The first person has a choice of 6 papers, so we put a 6 in the first box |6| | | |. Now there are 5 papers left. The second person has a choice of 5 papers, so we put a 5 in the second box |6|5| | |. We continue on until we get |6|5|4|3| . Thus, the total number of permutations that we can get of 6 things taken 4 at a time is $6 \times 5 \times 4 \times 3 = 360$

EXAMPLE 5: In an office there are 10 empty desks in a row. In how many ways can 7 new people be assigned to these 10 desks? In this problem 7 people have a choice of 10 desks. Therefore, the boxes will represent the 7 people | | | | | | | |. The first person has a choice of any of the 10 desks, so we put a 10 in the first box |10| | | | | | |. There are 9 desks left. The second person picks. He has a choice of any one of the 9 remaining desks, so we put a 9 in the second box |10|9| | | | | |. We can see that we will end up with |10|9|8|7|6|5|4|. Therefore, the permutations of 10 things taken 7 at a time is $10 \times 9 \times 8 \times 7 \times 6 \times 5 \times 4 = 604{,}800$

EXAMPLE 6: In how many ways can 5 items be deposited in 7 boxes if each item must be deposited in a different box? The 5 items have a choice of 7 boxes. Therefore, our answer is |7|6|5|4|3| = 2520 permutations

EXAMPLE 7: In how many ways can 8 horses finish in a race? There are 8 horses so we will have 8 boxes | | | | | | | | |. Horse #1 can finish 1st, 2nd, 3rd, 4th, 5th, 6th, 7th, 8th so we put an 8 in the first box |8| | | | | | | |. Horse #1 has finished in a certain position. Now horse #2 has to be accounted for. It can finish in any one of 7 positions, so we put a 7 in the second box |8|7| | | | | | |. If we continue on we will end up with |8|7|6|5|4|3|2|1| = 40320

EXAMPLE 8: How many code numbers can be made using 2 letters followed by a three-digit number? There are 5 spaces to fill | | | | | |. The first space can be filled with any one of 26 letters. There are 26 letters in the alphabet, so we put a 26 in the first box |26| | | | |. The second box can also be filled with any one of the 26 letters since repetitions of a letter are allowed. So we put a 26 in the second box |26|26| | | |. The third space can be filled in 10 ways since any of the one digit numbers (0 - 9) can be picked. So we put a 10 in the third box |26|26|10| | |. The fourth space can be filled in 10 ways, so we put a 10 in the 4th box |26|26|10|10| | . The fifth space can also be filled in 10 ways so we put a 10 in the 5th box |26|26|10|10|10|. The answer is: $26 \times 26 \times 10 \times 10 \times 10 =$ 676,000 numbers

EXAMPLE 9: How many code numbers can be made using 2 letters, followed by a three-digit number, which in turn is followed by 1 letter, if repetitions of letters and numbers are not allowed? There are 6 spaces to fill | | | | | | |. The first space can contain any one of the 26 letters in the alphabet so we put a 26 in the first box |26| | | | | |. One letter of the 26 has been used in space #1. There are 25 letters left to pick from since repetitions of a letter are not allowed. Therefore, we put a 25 in the second box |26|25| | | | | . The third space involves a number. This space can be filled in 10 ways since any of the one digit numbers 0 - 9 can be picked |26|25|10| | | |. One of the ten numbers has been used so there are 9 left. Thus, we put a 9 in box #4 |26|25|10|9| | |. Two numbers have been used so there are now 8 left, so we put an 8 in the 5th box |26|25|10|9|8| |. The last box involves a letter. Since repetitions of a letter are not allowed, we must put a 24 in the last box |26|25|10|9|8|24|. Therefore our answer is: $26 \times 25 \times 10 \times 9 \times 8 \times 24 = 11{,}232{,}000$ numbers

EXAMPLE 10: How many permutations can we get from the letters W, X, Y and Z, if we take them two at a time? There are going to be two boxes to fill. The first box has a choice of any one of the four letters, so we put a 4 in the first box |4| |. One letter has been used, so we put a 3 in the second box [4|3]. The answer is 4 × 3 = 12. The permutations are: WX, WY, WZ, XW, XY, XZ, YX, YW, YZ, ZW, ZX and ZY

EXAMPLE 11: In a special election, five men are running for dogcatcher and three men are running for town clerk. In how many different ways can the ballots be marked for these two offices? Note: the people running for town clerk cannot be elected as dog catcher and vice versa. There are two offices; therefore, there will be two boxes. Box #1 represents the dogcatcher. This box has a choice of any one of five people, so we put a 5 in it [5|]. The second box represents the town clerk so we put a 3 in it [5|3]. The ballot can thus be marked in 5 × 3 = 15 ways.

EXAMPLE 12: How many words can be formed from the letters of the word "DOG" taken all at a time? There are 3 letters, so we are going to have 3 boxes [| |]. The first box has a choice of any of the 3 letters so we put in it [3| |]. One letter has been used up! Now there are two left so we put a 2 in the second box [3|2|]. There is one left, so we put a 1 in the third box [3|2|1]. The answer is 3 × 2 × 1 = 6 words. They are DOG, DGO, GOD, GDO, OGD and ODG

EXAMPLE 13: In planning a round trip from New York to Los Angeles via Chicago, a salesman wishes to travel between New York and Chicago by train and between Chicago and Los Angeles by air.

If there are two railroad lines between New York and Chicago, and five airlines between Chicago and Los Angeles, in how many ways can the round trip be made if the salesman does not want to travel on the same line twice?

He can travel from New York to Chicago in two ways and from Chicago to Los Angeles in five ways. Since we now have used up one railroad and one airline, he can travel back from Los Angeles to Chicago in four ways and from Chicago to New York in one way. Using the box method we get [2|5|4|1]. The sales-man can, therefore, make this trip in 2 × 5 × 4 × 1 = 40 ways.

EXAMPLE 14: How many <u>even</u> three-digit numbers can we get from the numbers 3, 4, 5, 6 and 7?

The three-digit number must be even, so it should end in 4 or 6, since these are the only two even numbers in the data. There are three boxes to fill [| |]. The last box is filled first because the last digit determines whether the three-digit number is odd or even. This box has a choice of two digits, the 4 or the 6, so we put 2 in the last box [| |2]. One number has been picked so there are now 4 numbers left. Box #1 picks, it has a choice of 4 numbers, so we put a 4 in this box [4| |2]. Now there are three numbers left, so we put a 3 in box #2 [4|3|2]. We get 4 × 3 × 2 = 24 three-digit numbers

EXAMPLE 15: A sports car manufacturer offers seven optional accessories with his car. In how many different ways can his sports car be equipped? There are seven accessories, so there will be seven boxes to fill [| | | | | |]. There are two choices for each accessory - yes or no. So we of the seven boxes [2|2|2|2|2|2|2]. The sports car can therefore be equipped in 2 × 2 × 2 × 2 × 2 × 2 = 128 ways. This includes the fully equipped car and the stripped car.

FORMULAS FOR PERMUTATIONS

1. <u>FORMULA FOR PERMUTATIONS OF A SINGLE EVENT HAPPENING</u>. In the previous section, we learned how to solve permutation problems by a technique called the box method. We could have also used a formula to solve these problems.

   The permutation formula is

$$P = \frac{N!}{(N-r)!}$$

   where there are N things taken r at a time.

   The sign ! means factorial and it is the product of all the consecutive integers from 1 to the number preceding the factorial sign. For example:

   7! = 7 × 6 × 5 × 4 × 3 × 2 × 1 = 5040

   5! = 5 × 4 × 3 × 2 × 1 = 120

   2! = 2 × 1 = 2

   0! = 1 (by definition)

2. SAMPLE FORMULA METHOD PERMUTATION PROBLEMS.

EXAMPLE 1: In how many ways can two books be chosen from five different books and arranged in two spaces on a shelf ?

Answer: (A) via box method:

$\boxed{5}\boxed{4} = 20$

(B) via formula:

$$P = \frac{N!}{(N-r)!}$$

N = 5, r = 2 (5 things taken 2 at a time)

$$P = \frac{5!}{(5-2)!} = \frac{5!}{3!} = \frac{5 \times 4 \times 3 \times 2 \times 1}{3 \times 2 \times 1}$$

$$= 5 \times 4 = 20$$

EXAMPLE 2: How many arrangements are there of four cards taken from a deck of fifty-two cards?

52 cards taken 4 at a time.

N = 52, r = 4

$$P = \frac{52!}{(52-4)!} = \frac{52!}{48!}$$

$$= \frac{52 \times 51 \times 50 \times 49 \times 48 \text{ etc.}}{48 \times 47 \times 46 \times 45 \text{ etc.}}$$

$$= 6{,}500{,}000$$

(Order counts 2, 3, 4, 5 of hearts is different than a 2, 3, 5, 4 of hearts)

EXAMPLE 3: In how many ways can the letters of the word "QUICK" be arranged?

5 letters taken 5 at a time.

N = 5, r = 5

$$P = \frac{5!}{(5-5)!} = \frac{5!}{0!} = \frac{5 \times 4 \times 3 \times 2 \times 1}{1}$$

$$= 120$$

EXAMPLE 4: In how many ways can six books designated L, M, N, O, P and Q be arranged on a shelf ?

6 books taken 6 at a time.

N = 6, r = 6

$$P = \frac{6!}{(6-6)!} = \frac{6!}{0!} = \frac{6 \times 5 \times 4 \times 3 \times 2 \times 1}{1}$$

$$= 720$$

3. FORMULAS FOR PERMUTATION OF SEVERAL EVENTS HAPPENING. There are two more formulas for finding permutations.

1. Addition Law. The second formula is often called the "either-or = addition" law. It is: "If two events are mutually exclusive (i.e., the outcome of one event has no effect on the outcome of the other event) and the first event can be done in A ways, and the second can be done in B ways, then either one event or the other can be done in A + B ways."

2. Multiplication Law. The third formula is often called the "and = multiplication" law. It is: "If two events are mutually inclusive (i.e., the outcome of one event affects the outcome of the other) and the first event can be done in A ways, and the second event can be done in B ways, then when the two events are done together they can be performed in A times B ways.

4. FORMULA FOR PERMUTATIONS OF OBJECTS THAT ARE NOT ALL DIFFERENT.

So far, we discussed only the methods of finding the permutations of objects that are different from each other. We did not discuss the permutation of objects that are alike. For example, the three letters X, Y and Z yield six three letter words: XYZ, XZY, YXZ, YZX, ZXY, and ZYX. On the other hand, the three letters X, X, and Y yield three letter words: XXY, XYX, and YXX.

The formula for finding the number of permutations of objects that are not all different from each other is:

$$P = \frac{N!}{N_1!N_2!N_3!}$$

N = total number of objects

N_1 = number of objects of one kind

N_2 = number of objects of another kind

N_3 = number of objects of still another kind, etc.

$N = N_1 + N_2 + N_3 + \ldots.$

5. SAMPLE PROBLEMS INVOLVING PERMUTATIONS OF OBJECTS THAT ARE NOT ALL DIFFERENT.

PROBLEM 1: In how many ways can all the letters of the word ASSESS be arranged?

Solution

$$P = \frac{N!}{N_1!N_2!N_3!}$$

N - 6 letters

N_1- number of A's = 1

N_2- number of S's = 4

N_3- number of E's = 1

$$P = \frac{6!}{1!\,1!\,4!} = \frac{6\times5\times4\times3\times2\times1}{1\times1\times4\times3\times2\times1}$$

= 30 permutations.

The 30 permutations of the word ASSESS are:

ASSESS	SSSSAE	SASSES
ESSASS	SSSSEA	SESSAS
AESSSS	ASSSSE	SSASSE
EASSSS	ESSSSA	SSESSA
SAESSS	ASESSS	SSSASE
SEASSS	ESASSS	SSSESA
SSAESS	ASSSES	SASSSE
SSEASS	ESSSAS	SESSSA
SSSAES	SASESS	SSASES
SSSEAS	SESASS	SSESAS

PROBLEM 2: In how many different ways can the following ten machines be arranged in a straight line in a shop: (1) lathe, (1) punch press, (2) drill presses, (3) welding machines and (3) milling machines?

Solution

$$P = \frac{N!}{N_1!N_2!N_3!N_4!N_5!}$$

N = 10 machines

N_1 = number of lathes = 1

N_2 = number of punch presses = 1

N_3 = number of drill presses = 2

N_4 = number of welding machines = 3

N_5 = number of milling machines = 3

$$P = \frac{10!}{1!\,1!\,2!\,3!\,3!} = \frac{10\times9\times8\times7\times6\times5\times4\times3\times2\times1}{1\times1\times2\times1\times3\times2\times1\times3\times2\times1}$$

= 50,400 ways

INTRODUCTION TO COMBINATIONS. A combination is defined as any arrangement of objects where the order does not count. For example, ABC is the same as CBA, WXYZ is the same as WXZY, etc.

1. <u>FORMULA FOR FINDING COMBINATIONS</u>. The formula for finding combinations is:

$$C = \frac{N!}{r!\,(N-r)!}$$

Where there are N things taken r at a time.

2. <u>SAMPLE COMBINATION PROBLEMS</u>.

PROBLEM 1: In how many ways can a man choose three gifts from five different articles?

Solution

The order is not important, so it's a combination problem:

$$C = \frac{N!}{r!\,(N-r)!}$$

N = 5

r = 3

$$C = \frac{5!}{3!\,(5-3)!} = \frac{5\times4\times3\times2\times1}{3\times2\times1\times2\times1}$$

= 10

If we designate the five gifts as a, b, c, d and e, then the ten combinations are:

abc	ade	bcd	cde
cbd	acd	bce	
abe	ace	bde	

PROBLEM 2: In how many ways can a committee of ten be selected from a group of 12 people?

Solution

The order in which the committee is selected is not important, so this is a combination problem:

$$C = \frac{N!}{r!\,(N-r)!}$$

N = 12

r = 10

$$C = \frac{12!}{10!\,(12-10)!} = 66$$

3. <u>FORMULA FOR COMBINATIONS OF SEVERAL EVENTS HAPPENING.</u> There are two more formulas for finding combinations. They are:

a. <u>Addition Law</u>. The second formula is often called the "either-or addition" law. It is: "If two events are mutually <u>exclusive</u>

(i.e., the outcome of one event has no effect on the outcome of the other event) and the first event can be done in A ways, and the second event can be done in B ways, then either one event or the other can be done in A+ B ways."

2. Multiplication Law. The third formula is often called the "and = multiplication" law. It is: "If two events are mutually inclusive (i.e., the outcome of one event affects the outcome of the other event) and the first event can be done in A ways, and the second can be done in B ways, then when the two events are done together, they can be performed in A times B ways."

SAMPLE PERMUTATION AND COMBINATION PROBLEMS INVOLVING MULTIPLE EVENTS HAPPENING

PROBLEM 1: In how many ways can an investigating committee of 4 be chosen from a group of 3 engineers and 5 accountants, if the committee must consist of 2 engineers and 2 accountants?

Sol\ution In selecting members of a committee the order in which they are picked does not matter. Therefore, this is a combination problem.

(1) The 2 engineers for the committee can be picked from the group of 3 engineers in:

$$C = \frac{3!}{2!\,(3-2)!} = \frac{3!}{2!\,1!} = 3 \text{ ways}$$

(N = 3, r = 2)

(2) The accountants can be picked from the group of 5 accountants in:

$$C = \frac{5!}{2!\,(5-2)!} = \frac{5!}{2!\,3!} = 10 \text{ ways}$$

(N = 5, r = 2)

(3) This problem involves the two events of engineers being picked and accountants being picked. Therefore, we are going to multiply the two combinations together to get the total number of ways that the committee can be formed.

C = (C engineers) × (C accountants)

C = (3) (10) = 30 ways

PROBLEM 2: Each day an inspector visits 5 different machines in the morning and 4 different assembly operations in the afternoon. He varies the order of the visits to keep the operators from knowing just when to expect him. In how many different orders can he visit the operators?

Solution

The order counts, so this is a permutation problem.:

(1) In the morning the inspector can visit the 5 machines in:

$$P = \frac{5!}{(5-5)!} = \frac{5!}{0!} \quad \text{where n = 5, and r = 5}$$

$$= \frac{5\times4\times3\times2\times1}{1} = 120 \text{ ways}$$

(2) In the afternoon the inspector can visit the 4 operators in.

$$P = \frac{4!}{(4-4)!} = 24 \text{ ways where N = 4, r = 4}$$

(3) The problem involves two events, the order of visiting the 5 machines and the order of visiting the 4 assembly operations. Therefore, we have to multiply the two permutations together to get the total number of orders that the 9 operators can be visited in:

P = (P morning) and (P afternoon)

P = (120) (24) = 2,880 orders

PROBLEM 3: If a shipment of automobiles to a dealer must consist of either 3 differently colored convertibles or 4 differently colored sedans, in how many ways can the shipment be picked from a group of 8 differently colored convertibles and 5 differently colored sedans?

Solution

The order in which the cars are picked out of the group is not important. Thus, this is a combination problem.

(1) The convertibles can be picked from the group of 8 in:

$$C = \frac{8!}{3!\,(8-3)!} = \frac{8!}{3!\,5!}$$

$$= \frac{8\times7\times6\times5\times4\times3\times2\times1}{3\times2\times1\times5\times4\times3\times2\times1}$$

C = 56 ways

(N = 8, r = 3)

(2) The 4 sedans can be picked from the group of 5 in:

$$C = \frac{5!}{4!\,(5-4)!} = \frac{5!}{4!\,1!}$$

= 5 ways

(N = 5, r = 4)

(3) This problem involves a shipment of either 3 convertibles or 4 sedans. Since it's either-or, we are going to add the two combinations together to get the number of ways that the shipment can be made.

C = (C convertibles) and (C sedans)

C = (56) + (5) = 61 ways

8 PROBABILITY

HOW TO CALCULATE THE PROBABILITY THAT EVENTS WILL HAPPEN

INTRODUCTION TO PROBABILITY. An ordinary deck of 52 playing cards contains 4 kings. If we pick one card at random from a well-shuffled deck there would be 4 chances out of 52 that the card we picked is a king. Similarly, a coin tossed in the air may land either heads or tails. Thus, since there are two possibilities, there is one chance out of two that the coin will land tails.

If we substitute the word probability for the word chance in the above example, then the probability of getting a king is 4 out of 52 and the probability of getting tails is one out of two.

1. DEFINITION OF PROBABILITY. Probability is a measure of the chance that an event will happen the way we want it to happen. The formula for the probability is:

$$\text{Probability} = \frac{\text{Number of Desired Outcomes}}{\text{Total Number of Outcomes}}$$

In our card example, the probability of drawing a king =

$$\frac{4}{52} = \frac{1}{13}$$

and,in the coin example the probability of getting tails = $\frac{1}{2}$

In mathematical terms, the probability of an event happening can have a value from 0 to 1. An event which is certain to happen will have a probability of 1 or 100%. An event which has no chance whatsoever of happening will have a probability of 0 or 0%.

2. HOW TO CALCULATE SIMPLE PROBABILITIES. The easiest way of illustrating the calculation of simple probabilities is through the solution of a few sample problems.

EXAMPLE 1: In drawing a card from a deck of 52 playing cards, what is the probability of getting a spade:

Solution There are 13 spades in a deck of cards. The desired outcome is drawing a spade.

Probability of getting a spade

$$= \frac{\text{Number of Desired Outcomes}}{\text{Total Number of Outcomes}} = \frac{13}{52} = \frac{1}{4}$$

(One out of every 4 cards is a spade.)

EXAMPLE 2: A bowl contains 10 blue balls, 20 yellow balls and 30 red balls. What is the probability of drawing a blue ball out of this bowl?

Solution Desired outcome is drawing a blue ball. The number of desired outcomes = (10 + 20 + 30) = 60

Probability of getting a blue ball

$$= \frac{10}{60} = \frac{1}{6}$$

(One out of every 6 balls is blue.)

EXAMPLE 3: In a 5,000 ticket drawing, what is a person's probability of losing if he buys 50 tickets?

Solution Desired event is losing. Number of desired outcomes = (5,000 – 50) = 4,950

The total number of outcomes = 5,000

$$\text{Probability of losing} = \frac{4{,}950}{5{,}000}$$

(4,950 out of 5,000 tickets are not owned by the individual)

INTRODUCTION OF THE PROBABILITY OF MULTIPLE EVENTS HAPPENING. In a previous section in this topic, we discussed a method for finding the probability that a single event will take place. To find the probability of a series of events happening, we use two theorems called the addition law and the multiplication law.

(1). ADDITION THEOREM FOR FINDING PROBABILITY. When two events are mutually exclusive, then the outcome of one event has no effect on the outcome of the other event. If the probability of the first event occurring is P(A) and the probability of the second event occurring is P(B), then the probability that either the first event or the second event will take place is P(A) + P(B).

This theorem is sometimes referred to as the "if it is either one event or the other, then add" law.

EXAMPLE: A bag contains 7 red balls, 5 blue balls and 3 white balls. What is the probability of drawing either a red or a white ball out of the bag?

Solution

$$P\ (\text{red}) = \frac{7}{7+5+3} = \frac{7}{15}$$

$$P\ (\text{white}) = \frac{3}{7+5+3} = \frac{3}{15}$$

Since the question is either red or white we add the two probabilities together:

P(red or white) = P(red) + P(white)

$$= \frac{7}{15} + \frac{3}{15} = \frac{10}{15} = \frac{2}{3}$$

(2). MULTIPLICATION THEOREM FOR FINDING PROBABILITY. When two events are mutually inclusive, then the outcome of one event affects the outcome of the other event. If the probability of the first event occurring is P(A) and the probability of the second event occurring is P(B), then the probability that the first event and the second event will both occur is P(A) × P(B).

This theorem is sometimes referred to as the "if it is one event and the other event, then multiply" law.

EXAMPLE: A bag contains 7 red balls, 5 blue balls and 3 white balls. What is the probability of drawing a red and a white ball out of the bag on two successive draws with replacement?

Solution

$$P\ (\text{red}) = \frac{7}{7+5+3} = \frac{7}{15}$$

$$P\ (\text{white}) = \frac{3}{7+5+3} = \frac{3}{15}$$

The question is red and white so we multiply the two probabilities together:

P(red and white) = P(red) × P(white)

$$= \frac{7}{15} \times \frac{3}{15} = \frac{21}{225}$$

3. SAMPLE PROBABILITY OF MULTIPLE EVENT PROBLEMS.

PROBLEM 1: A bowl contains 20 red balls, 10 green balls, 5 white balls and 15 blue balls. What is the probability of drawing either a red, green, white or blue ball out of the bowl?

Solution Total number of balls

= (20 + 10 + 15 + 5) = 50

$$P(\text{red}) = \frac{20}{50}$$

$$P(\text{green}) = \frac{10}{50}$$

$$P(\text{white}) = \frac{5}{50}$$

$$P(\text{blue}) = \frac{15}{50}$$

P(red or green or white or blue)

$$= \left(\frac{20}{50} + \frac{10}{50} + \frac{5}{50} + \frac{15}{50}\right) = \frac{50}{50} = 1$$

One, you remember, means absolute certainty the ball picked has to be one of the four colors. Thus, the total probability of all the possible outcomes = 1 = 100%.

PROBLEM 2: What is the probability of throwing 3 fives in a row with one die? (The plural of die is dice.)

Solution A die has six sides, each of the numbers one, two, three, four, five, and six have an equal chance of being on top. Therefore, the probability of getting a five is:

$$P(5) = \frac{1}{6}$$

The probability of throwing three 5's in a row is the probability of throwing a 5, another 5 and still another 5 in a row. Since it is 5 and 5 and 5 we use the multiplication theorem:

P(three 5's) = P(5) × P(5) × P(5)

$$= \left(\frac{1}{6}\right) \times \left(\frac{1}{6}\right) \times \left(\frac{1}{6}\right) = \frac{1}{216}$$

PROBLEM 3: What is the probability of getting either a diamond or a heart in a single draw from a standard deck of 52 playing cards?

Solution The question is either diamond or heart, so we use the addition theorem. A deck contains 13 diamonds, 13 hearts, 13 spades and 13 clubs:

$$P(\text{diamonds}) = \frac{13}{52} = \frac{1}{4}$$

$$P(\text{heart}) = \frac{13}{52} = \frac{1}{4}$$

$$P(\text{diamonds or hearts}) = \frac{1}{4} + \frac{1}{4} = \frac{1}{2}$$

PROBLEM 4: If a die is tossed twice, what is the probability of getting a two on the first toss followed by a four on the second toss?

Solution A die has 6 sides. Each number has an equal chance of being on top. The probability of getting a two <u>and</u> a four involves the multiplication theorem:

$$P(2) = \frac{1}{6}$$

$$P(4) = \frac{1}{6}$$

$$P(2 \text{ and } 4) = \left(\frac{1}{6} \times \frac{1}{6}\right) = \frac{1}{36}$$

PROBLEM 5: A bag contains 20 blue balls and 30 white balls. What is the probability of drawing two successive blue balls out of this bag if the first ball is not replaced in the bag?

Solution The probability of drawing a blue ball <u>and</u> a blue ball involves the multiplication theorem.

To satisfy the question, the first ball must be blue. Therefore, since there is no replacement there will be only 19 blue balls left in the bag for the second draw:

<u>1st draw</u> $P(\text{blue}) = \frac{20}{50}$

<u>2nd draw</u> $P(\text{blue}) = \frac{19}{49}$

$$P(\text{blue and blue}) = \left(\frac{20}{50}\right)\left(\frac{19}{49}\right) = \frac{380}{2450}$$

$$= \frac{38}{245}$$

PROBLEM 6: Through an error 450 defective radio tubes were randomly mixed in a total inventory of 45,000 tubes. How many defective tubes could be expected in a shipment of 1,000 tubes out of this inventory?

Solution Probability of a tube being defective

$$= \frac{450}{45{,}000} = \frac{1}{100} = .01$$

One out of every 100 tubes is defective. Thus in a shipment of 1000 tubes, we can expect to have (.01) (1000) = 10 defective tubes.

PROBLEM 7: In three successive draws from an ordinary deck of 52 playing cards, what is the probability of getting three aces if the cards are not replaced after each draw?

Solution A deck of cards contains 4 aces. The first card must be an ace, so after the 1st draw there will be 3 aces left. The second card too must be an ace, so after the second draw there will be 2 aces left. The third card must be an ace, so after the 3rd draw there will be 1 ace left. The drawing will be reduced by one card after each draw.:

<u>1st draw</u> $P(\text{ace}) = \frac{4}{52}$

<u>2nd draw</u> $P(\text{ace}) = \frac{3}{51}$

<u>3rd draw</u> $P(\text{ace}) = \frac{2}{50}$

Probability of getting an ace on the first <u>and</u> second <u>and</u> third draw is:

$$P(\text{three aces in a row}) = \frac{4}{52} \times \frac{3}{51} \times \frac{2}{50}$$

$$= \frac{24}{132{,}600}$$

PROBLEM 8: A bag contains 5 blue balls and 10 red balls. In two random draws from this bag, what is the probability of getting one blue ball and one red ball with replacement?

Solution

$$P(\text{blue ball}) = \frac{5}{15} = \frac{1}{3}$$

$$P(\text{red ball}) = \frac{10}{15} = \frac{2}{3}$$

In two random draws, the first ball could be blue and the second red <u>or</u> the first could be red and the second blue and still give us one red and one blue ball:

$$P(\text{blue and red}) = \left(\frac{1}{3} \times \frac{2}{3}\right) = \frac{2}{9}$$

$$P(\text{red and blue}) = \left(\frac{2}{3} \times \frac{1}{3}\right) = \frac{2}{9}$$

Probability of getting <u>either</u> a blue and <u>red</u> <u>or</u> a red and blue ball is:

$$P(\text{blue, red or red, blue}) = \left(\frac{2}{9} + \frac{2}{9}\right) = \frac{4}{9}$$

PROBLEM 9: Construct a table showing all the possible outcomes that can result from tossing two ordinary dice, and determine the probability of:

(A) Throwing a double;

(B) getting a 5, and

(C) getting either a 7 or an 11

Solution The following table shows all the outcomes for a two-die toss:

NUMBER APPEARING ON SECOND DIE

		1	2	3	4	5	6
Number Appearing on First Die	1	1, 1	1, 2	1, 3	1, 4	1, 5	1, 6
	2	2, 1	2, 2	2, 3	2, 4	2, 5	2, 6
	3	3, 1	3, 2	3, 3	3, 4	3, 5	3, 6
	4	4, 1	4, 2	4, 3	4, 4	4, 5	4, 6
	5	5, 1	5, 2	5, 3	5, 4	5, 5	5, 6
	6	6, 1	6, 2	6, 3	6, 4	6, 5	6, 6

We can see from the table that there are 36 possible outcomes for a two dice toss:

(A) The probability of throwing a double is found by adding up all the doubles in the table and then dividing by the total number of outcomes, which is 36:

There are 6 doubles: (1, 1) (2, 2) (3, 3) (4, 4) (5, 5) and (6, 6).

$$P\text{ (double)} = \frac{6}{36} = \frac{1}{6}$$

(B) The probability of getting a 5 is found by counting the number of outcomes that add up to 5. (1, 4) (2, 3) (3, 2) and (4, 1) are all the outcomes that add up to 5:

$$P(5) = \frac{4}{36} = \frac{1}{9}$$

(C) The probability of getting <u>either</u> a 7 <u>or</u> an 11 is equal to the probability of getting a 7 plus the probability of getting an 11:

The outcomes that add up to 7 are;

(1, 6) (2, 5) (3, 4) (4, 3) (5, 2) and (6, 1).

There are 6 outcomes that add up to 7:

$$P\text{ (7)} = \frac{6}{36}$$

The two outcomes that add up to 11 are: (5, 6) and (6, 5).

$$P\text{ (11)} = \frac{2}{36}$$

Probability of getting either a 7 or an 11 is:

$$= \frac{6}{36} + \frac{2}{36} = \frac{8}{36} = \frac{2}{9}$$

PROBABILITY OF AN EVENT NOT HAPPENING. The probability of an event happening is given by the formula:

$$\text{Probability} = \frac{\text{Number of Desired Outcomes}}{\text{Total Number of Outcomes}}$$

The total probability of <u>all</u> outcomes is equal to 1 or 100% because one of all the outcomes has to occur. Therefore, the probability of an event <u>not</u> happening in your favor is equal to one minus the:

$$1 - \frac{\text{(Probability of the event happening in your favor, or minus the Number of Outcomes)}}{\text{(Total Number of Outcomes)}}$$

EXAMPLE: In drawing a card from an ordinary deck of 52 playing cards what is the probability of not drawing a 10?

Solution There are four 10's in a deck of cards. Probability of drawing a 10

$$= \frac{4}{52}$$

Probability of not drawing a 10

$$= \left(1 - \frac{4}{52}\right) = \frac{48}{52}$$

(48 of the 52 cards are not 10's.)

9 BINOMIAL DISTRIBUTION

HOW TO USE THE BINOMIAL DISTRIBUTION TO COMPUTE PROBABILITIES

INTRODUCTION TO THE BINOMIAL DISTRIBUTION. The binomial distribution is a method for determining the probabilities of all the different ways that an event can occur. For example, if a large bag contains red balls and white balls, and we take a sample of 4 balls from this bag, the sample can contain either 0, 1, 2, 3, or 4 red balls. By using terms in the binomial expansion, we can find the probabilities of getting any one of the five combinations of red balls in the sample. In other words, the binomial can be used to determine the probability of getting no red balls in the sample, the probability of getting one red ball in the sample, the probability of getting two red balls in the sample, etc.

THE BINOMIAL EXPANSION. A binomial expression contains two terms joined by a plus or minus sign. Examples of binomial expressions are:

$(a + b)^2$, $(2X + 7Y)^8$, $(p + q)^2$, etc.

The general formula for the expansion of a binomial expression is:

$$(a+b)^N = a^N + \frac{Na^{N-1}b}{1!} + \frac{N(N-1)a^{N-2}b^2}{2!}$$

$$+ \frac{N(N-1)(N-2)a^{N-3}b^3}{3!} + \ldots\ldots\ldots$$

N = the exponent

a = the first term in the binomial

b = the second term in the binomial

! = factorial sign. It is the product of all the consecutive numbers from 1 to the number preceding the factorial sign. For example,

$5! = 5 \times 4 \times 3 \times 2 \times 1 = 120$

$3! = 3 \times 2 \times 1 = 6$

$0! = 1$ (by definition)

The expansion stops when the exponent power for a is equal to zero. This is where a drops out of the equation because any number to the zero power is always equal to 1. For example, $7^0 = 1{,}1000^0 = 1$. The last term thus becomes b^N.

EXAMPLE 1: Use the general formula to expand

$(p + q)^3$

Solution

$a = p, b = q, N = 3$

$$(p+q)^3 = p^3 + \frac{3p^2q}{1!} + \frac{3(2)pq^2}{2!} + \frac{3(2)(1)p^0q^3}{3!}$$

$$(p + q)^3 = p^3 + 3p^2q + 3pq^2 + q^3$$

EXAMPLE 2: Use the general formula to expand the binomial $(p + q)^2$:

Solution

$a = p, b = q, N = 2$

$$(p+q)^2 = p^2 + \frac{2pq^1}{1!} + \frac{2(1)p^0q^2}{2!}$$

$$(p + q)^2 = P^2 + 2pq + q^2$$

THE BINOMIAL DISTRIBUTION. The probability that an event will either happen or will not happen = 1. There are no other alternatives. If we let p = the probability that an event will happen and q = the probability that an event will not happen, then (p + q) = 1.

If a box contains a large number of beads in which 20% are colored green and the rest are white, then the probability that a single bead, chosen at random from this box is colored green = p = 20% = .2. The probability that a single bead is colored white is q = 80% = .8. The probability that a single bead is either green or white is = (p + q) = (.2 + .8) = 1.

If we now randomly pick two beads, one at a time from this box, they can both be green, both be white. The first bead picked green and the second white, or the first bead picked white and the second green. There are three results and four ways to getting them.

The probability that both beads are green	$= p \times p = p^2 = (.2)(.2) = .04$
The probability that both beads are white	$= q \times q = q^2 = (.8)(.8) = .64$
The probability that the first is green and second is white	$= p \times q = pq = (.2)(.8$ $= .16$
The probability that the first is white and the second is green	$= q \times p = qp = (.2)(.8)$ $= .16$

The total probability for all the combinations is equal to their sum:

$$= p^2 + 2pq + q^2 = (.04 + .32 + .64) = 1.0$$

$p^2 + 2pq + q^3$ is equal to the binomial expansion of the expression $(p + q)^2$, where the exponent 2 is equal to the sample size. Thus, we can use successive terms of the binomial expansion to find the probability of the different combinations of ways that an event can occur.

If we now pick three beads one at a time from this box, we could get the following results:

Result	Way of Getting Result			Probability	Total Probability for Result
	1st Bead	2nd Bead	3rd Bead		
3 green	green	green	green	$p.p.p = p^3$	p^3
2 green and 1 white	green green white	green white green	white green green	$p.p.q = p^2q$ $p.q.p = p^2q$ $q.p.p = p^2q$	$3(p^2q)$
1 green and 2 white	green white white	white green white	white white green	$p.q.q = pq^2$ $q.p.q = pq^2$ $q.q.p = pq^2$	$3(pq^2)$
3 white	white	white	white	$q.q.q = q^3$	q^3

There are four results and eight ways of getting them. The total probability for the four possible results is equal to the sum of the right hand column $= p^3 + 3p^2q + 3pq^2 + q^3 = (.2)^3 + 3(.2)^2(.8) + 3(.2)(.8)^2 + (.8)^3 = (.008 + .096 + .384 + .512) = 1.$

$p^3 + 3p^2q + 3pq^2 + q^3$ is equal to the binomial expansion of the expression $(p + q)^3$, where the exponent 3 is equal to the sample size. Again we've shown that the binomial expansion can be used to find the probability of getting different results.

THE GENERAL BINOMIAL FORMULA. We can use a general formula to find the probability of the specific result we want without using a table and without expanding the binomial. The general formula is:

$$P(X) = \frac{N!}{X!\,(N-X)!}\, p^X q^{N-X}$$

N = the sample size

x = result desired (for example 2 green beads in the sample)

p = probability that the desired event (x) will occur

q = probability that an event will not occur = (1 - p).

EXAMPLE: If 20% of the beads in a large box are colored green, what is the probability of getting 2 green beads in a sample of three?

Solution

$$P(X) = \frac{N!}{X!\,(N-X)!}\, p^X q^{N-X}$$

P(x) = probability of getting 2 green beads

X = 2 green beads

p = desired result = getting green beads = 20% = .2

q = (1 – p) = (1 – .2) = .8

N = sample size = 3

$$P(2) = \frac{3!}{2!(3-2)!}(.2)^2(.8)^1$$

$$= \frac{3\bullet 2\bullet 1}{2\bullet 1\bullet 1}(0.4)(.8)$$

$$P(2) = 3(.04)(.8) = .096$$

The probability of getting 2 green beads in a sample of three is = .096. This is the same result that we got before in the table.

1. SAMPLE BINOMIAL PROBLEMS.

PROBLEM 1: A sample of four items is taken at random from a production process, in which 10% of the items produced are defective. What is the probability of getting 2 defectives in the samples?

Solution

$$P(X) = \frac{N!}{X!(N-X)!}p^X q^{N-X}$$

N = 4, p = .1, q = .9, X = 2

$$P(2) = \frac{4!}{2!\,(4-2)!}(.1)^2(.9)^2$$

$$P(2) = \frac{4\bullet 3\bullet 2\bullet 1}{2\bullet 1\bullet 2\bullet 1}(.01)(.81)$$

$$P(2) = 6(.0081) = .0486$$

(About 48 out of every 1000 samples of 4 should have 2 defectives in the sample.)

PROBLEM 2: A sample of four items is taken at random from a production process in which 10% of the items produced are defective. What is the probability of getting two good units in the sample?

Solution

$$P(X) = \frac{N!}{X!\,(N-X)!} p^X q^{N-X}$$

N = 4, p = .9, q = 1,X = 2

p = probability of the desired result. In this case the desired result is getting good units:

p = 90% = .9

q = (1 − p) = (1 − .9) = .1

X = 2

$$P(2) = \frac{4!}{2!\,(4-2)!} (.9)^2 (.1)^2$$

= 6 (.81) (.01) = .0486

The probability of getting 2 good units in the sample of four is the same as the probability of getting two bad units in the sample. We can verify this by checking the previous problem.

PROBLEM 3: In a lot of electric bulbs, 30% of the bulbs are defective. What is the probability of getting three defective bulbs in a sample of five?

Solution

$$P(X) = \frac{N!}{X!\,(N-X)!} p^X q^{N-X}$$

N = 5, p = .30, q = .70, X = 3

$$P(3) = \frac{5!}{3!(5-3)!} (.3)^3\ (.7)^2$$

$$P(3) = \frac{5 \bullet 4 \bullet 3 \bullet 2 \bullet 1}{3 \bullet 2 \bullet 1 \bullet 2 \bullet 1} (.3)^3\ (.7)^2$$

P(3) = 10 (.027) (.49) = .1323

PROBLEM 4: If a soldier averages seven hits out of every ten shots at a target, what is the probability that he will hit the target in two out of three shots?

Solution

$$P(X) = \frac{N!}{X!\,(N-X)!} p^X q^{N-X}$$

p = event we want to happen = probability of hitting a target = 7 out of 10 = 70% = .7

q = (1 − p) = (1 − .7) = .3

N = sample size = 3 shots

X = desired event = 2 hits

$$P(2) = \frac{3!}{2!\,(3-2)!} (.7)^2\ (.3)^1$$

= 3 (.49) (.3) = .441

(There is a 44% chance that he will hit the target in 2 out of 3 shots.)

PROBLEM 5: If 40% of the children born in a certain hospital are girls, what is the probability that out of 5 children born on a certain day two of them will be girls?

Solution

$$P(X) = \frac{N!}{X!\,(N-X)!} p^X q^{N-X}$$

p = event we want = probability of girls = 40% = .4

q = (1 − p) = (1 − .4) = .6

X = 2

N = 5

$$P(4) = \frac{5!}{2!\ (5-2)!} (.4)^2\ (.6)^3$$

P(4) = 10 (.16) (.216) = .3456

PROBLEM 6: What is the probability of getting 3 heads and 2 tails in 5 flips of a coin?

Solution

$$P(X) = \frac{N!}{X!\,(N-X)!} p^X q^{N-X}$$

p = probability of getting heads $= \frac{1}{2} = .5$

q = probability of getting tails $= \frac{1}{2} = .5$

x = 3 heads

N = 5

$$P = \frac{5!}{3!\,(5-3)!} (.5)^3\ (.5)^2$$

$$P = \frac{5!}{3!\,2!} (.125)\,(.25)$$

P = 10 (.125) (.25) = .3125

PROBLEM 7: What is the probability of getting exactly one 6 in 4 rolls of a single die?

Solution

$$P(X) = \frac{N!}{X!(N-X)!} p^X q^{N-X}$$

p = probability of getting a 6 in 1 roll of a die = 1/6

q = (1 - p) = (1 - 1/6) = 5/6

x = 1

N = 4 rolls

$$P = \frac{4!}{1!(4-1)!}(1/6)^1 \ (5/6)^3$$

$$= 4(1/6)\frac{125}{216} = \frac{500}{1296} = .386$$

PROBLEM 8: If the probability that a person responds to a direct mail advertisement is .10, what is the probability of getting either 2, 3, or 4 responses to an ad which is mailed to 6 people?

Solution

Step 1. Probability of getting 2 responses is:

$$P(X) = \frac{N!}{X!(N-X)!} p^X q^{N-X}$$

p = probability of getting a response = .1

q = (1 – p) = (1 –.1) = .9

X = 2 responses

N = 6 people

$$P = \frac{6!}{2!\,(6-2)!} (.1)^2 \ (.9)^4$$

= 15 (.01) (.6561) = .0984

Step 2. Probability of getting 3 responses is:

$$P(X) = \frac{N!}{X!(N-X)!} p^2 q^{N-1}$$

p = .1

q = .9

x = 3 responses

N = 6

$$P = \frac{6!}{3!(6-3)!} (.1)^3 \ (.9)^3 = .0146$$

Step 3. Probability of getting 4 responses is:

$$P(X) = \frac{N!}{X!(N-X)!} p^X q^{N-X}$$

p = .1

q = .9

x = 4 responses

N = 6

$$P = \frac{6!}{4!\,(6-4)!} (.1)^4 \ (.9)^2$$

P = 15 (.0001) (.81) = .001215

Step 4. Probability of getting either 2,3 or 4 responses is equal to the sum of the individual probabilities:

$P = P_2 + P_3 + P_4$

P = (.0985 + .0146 + .001215)

P = .11422 or about .114

10 NORMAL DISTRIBUTION

HOW TO USE THE NORMAL DISTRIBUTION TO COMPUTE PROBABILITIES

INTRODUCTION TO THE NORMAL CURVE. If a frequency histogram, i.e., a bar graph of frequency versus class interval, were plotted for a very large amount of data, the graph would look something like this:

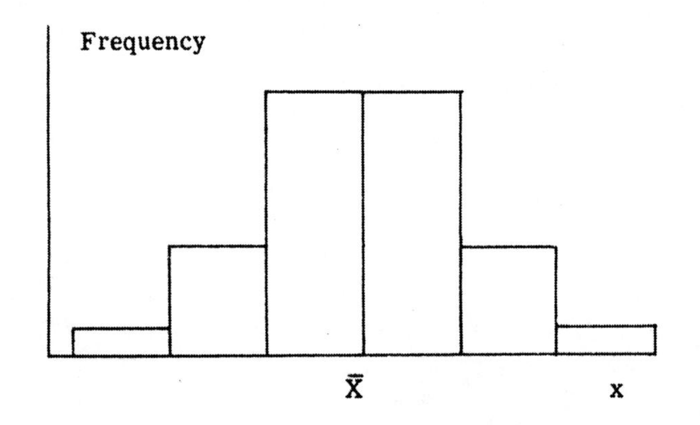

Now if the middle of the tops of each column were connected with a smooth curve, the curve would have a bell shape as shown below:

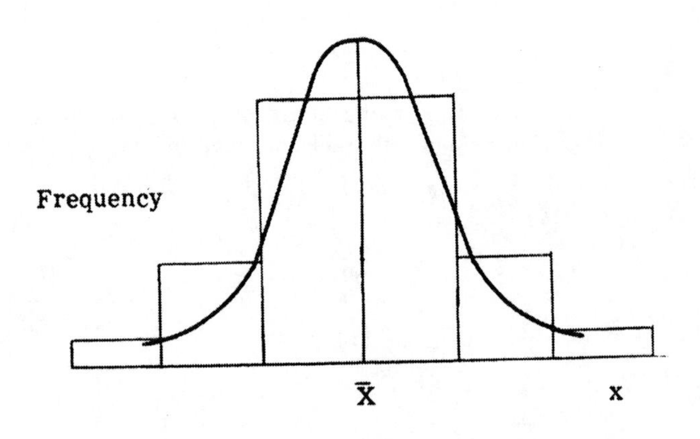

A perfect bell shaped curve is called a normal curve and the data represented by a normal curve is called a normal distribution.

Many distributions of data follow the pattern of the normal distribution. It is a very common or natural distribution found almost everywhere. That is why it is called "normal." We would, for example, probably obtain normal distributions for data plotted on such things as the IQ's of thousands of people, or the height of a great many male college freshmen, or the annual incomes of thousands of accountants, or the variations in a dimension on a part which is produced in large numbers, or the life of a certain make automobile tire, or the weights of thousands of apples, etc.

AREAS OF THE NORMAL DISTRIBUTION. The normal curve is symmetrical about its midpoint which is the MEAN of the data. Thus one-half of the area of this curve is left of the MEAN and one-half of the area is to the right of the MEAN as shown below:

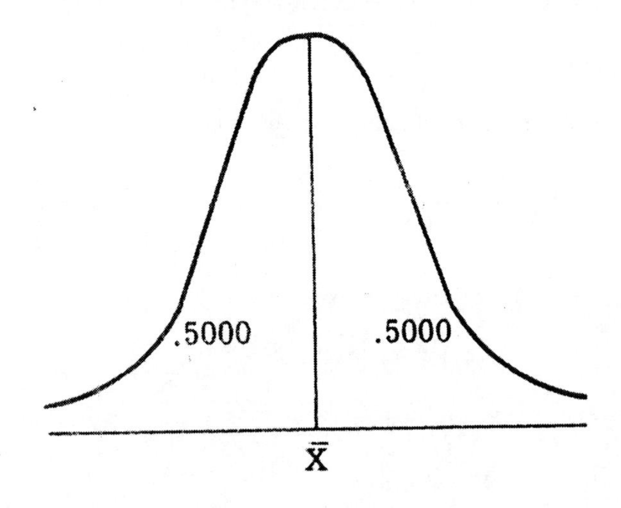

The normal curve is shaped so that 68.26% of its area lies within a ± 1 standard deviation distance from the MEAN. 95.44% of its area lies within a ± 2 standard deviation distance from the MEAN. Thus, if for example the MEAN of a normal distribution is 200 and the standard deviation is 10, we know that about 68% of the data lies between the values 190 and 210 ($200 \pm 1\sigma$, where $1\sigma = 10$). About 95% of the data lies between the values 180 and 220 ($200 \pm 2\sigma$, where $2\sigma = 20$), and almost all of the data lies between the value 170 and 230 ($200 \pm 3\sigma$, where $3\sigma = 30$. This example is illustrated on the following page:

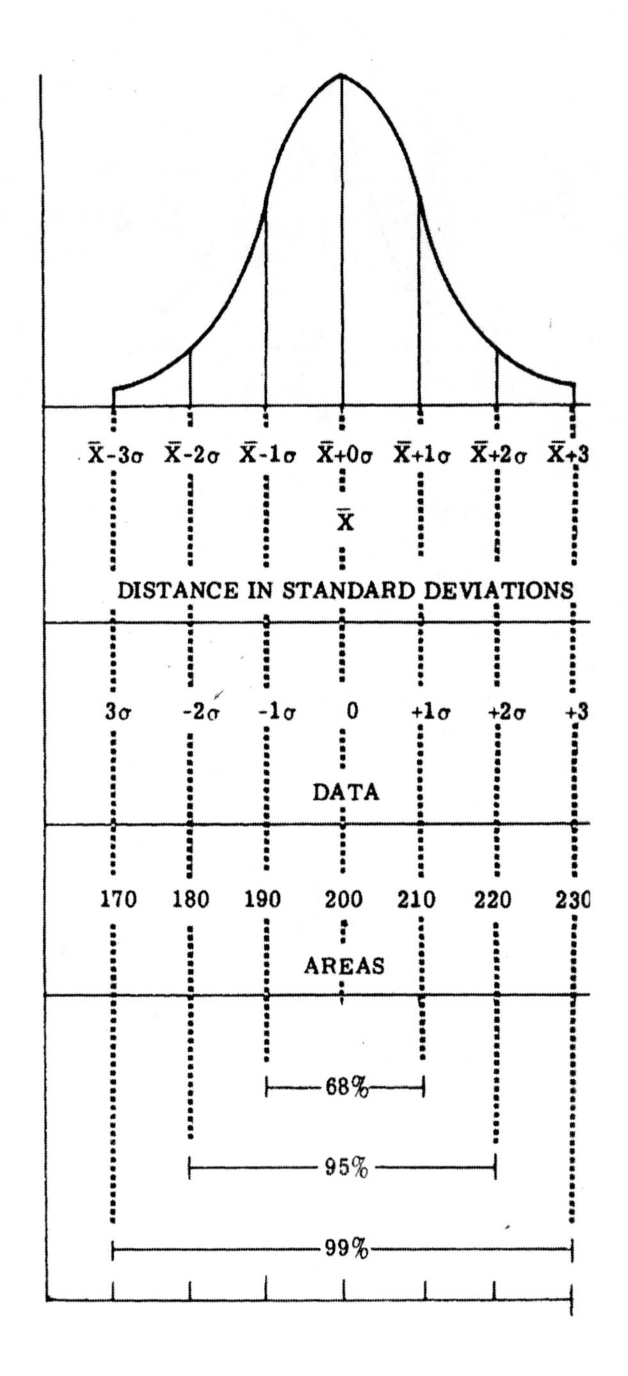

The standard deviation represents the average difference, in the data, from the MEAN value. Therefore the standard deviation equals zero at the MEAN. We can relate the areas under the normal curve to the standard deviation distance of a value from the MEAN, by using the formula:

$$Z = \frac{X - \overline{X}}{\sigma}$$

$X - \overline{X}$ = difference between a value X and the MEAN $\overline{X}$.

$\frac{X - \overline{X}}{\sigma}$ = the number of standard deviations that a value of X is away from the MEAN. It is the number of σ's in $X - \overline{X}$

EXAMPLE 1: If the MEAN of a normal distribution is 300 and the standard deviations is 15, find how many standard deviations away the number 450 is from the MEAN:

Solution

$$Z = \frac{X - \overline{X}}{\sigma} = \frac{450 - 300}{15} = \frac{150}{15} = 10$$

The value 450 is 10 standard deviations away from the MEAN. As a check, 10 standard deviations represent 10 × 15 = 150. $\overline{X}$ = 300, X therefore = 300 + 150 = 450 which is our original number.

EXAMPLE 2: If the MEAN of a normal distribution is 50 and the standard deviation is 25, find how many standard deviations away the number 25 is from the MEAN:

Solution

$$Z = \frac{X - \overline{X}}{\sigma} = \frac{25 - 50}{25} = \frac{-25}{25} = -1$$

The value 25 is 1 standard deviation away from the MEAN. The minus sign means that the value 25 is less than the value of the MEAN. Minus indicates left of the center line.

THE TABLE OF AREAS UNDER THE NORMAL CURVE. If we let the total area under the normal curve equal 1 or 100%, we can use the areas under the curve to represent relative frequencies or probabilities. Thus, the area under the curve between any two values of Z would represent the probability that any number, chosen at random from the data, will fall between the values of X which correspond to the two values of Z.

The following table contains the percentage of the areas under one half of a normal curve for various values of Z. All areas are measured from the center line of the curve where:

$Z = 0$

Only one-half of the areas are tabulated because the curve is symmetrical about its center line. The areas in the table therefore, only go up to .5000 or 1/2. The row is the first decimal place in Z and the column is the second decimal place. Thus for example, the Z = 1.45. We look in the 1.4 row and the .05 column to get area = .4265 at Z = 1.45

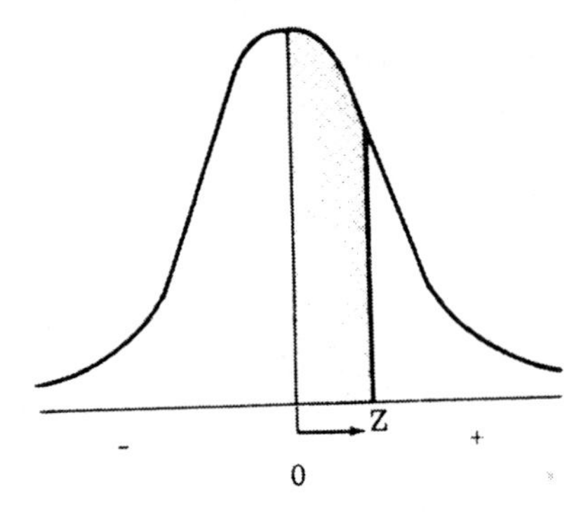

PROCEDURE FOR USING THE NORMAL DISTRIBUTION TO COMPUTE PROBABILITIES. The procedure for using the areas under a normal curve to compute probabilities will be illustrated through the solution of a sample problem.

EXAMPLE: The lifetime of a certain make of automobile headlight has a MEAN of 500 burning hours and a standard deviation of 60 hours. Find (A) the percentage of lights which have a lifetime greater than 380 hours, (B) the life above which we will find the best 20% of the headlights, and (C) the percentage of the headlights which have a life between 440 and 530 hours.

Solution (A):

Step 1. Draw a normal curve, label the MEAN, the value of X, and shade the desired area:

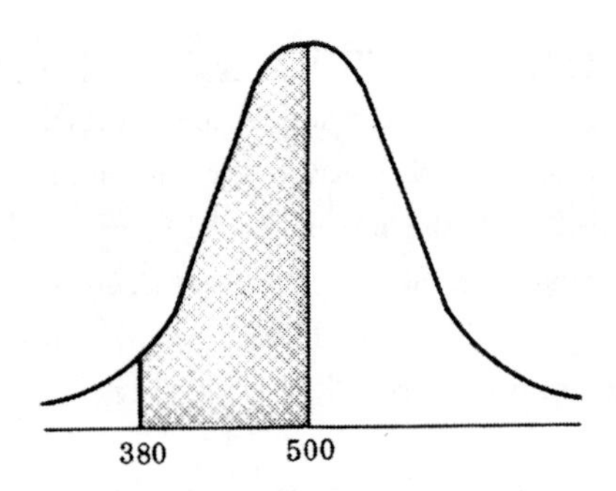

Step 2. Convert the value of X to its Z equivalent:

$$Z = \frac{X - \overline{X}}{\sigma} = \frac{380 - 500}{60} = \frac{-120}{60} = -2$$

The number 380 is 2 standard deviations to the left of the MEAN. The minus signs mean less than or to the left of the MEAN.

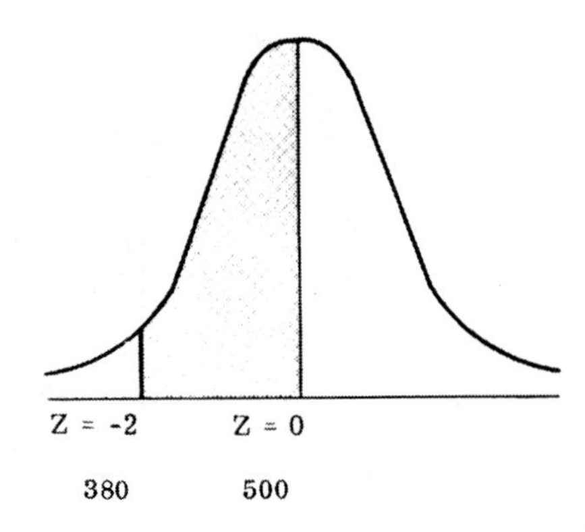

Step 3. Look up the area for the value of Z = 2 in the table of normal curve areas. Ignore the sign of Z. (Look up 2.0 row, .00 column in the table):

At Z = 2.00, area = .4772

This is the area between centerline where Z = 0 to the point Z = −2

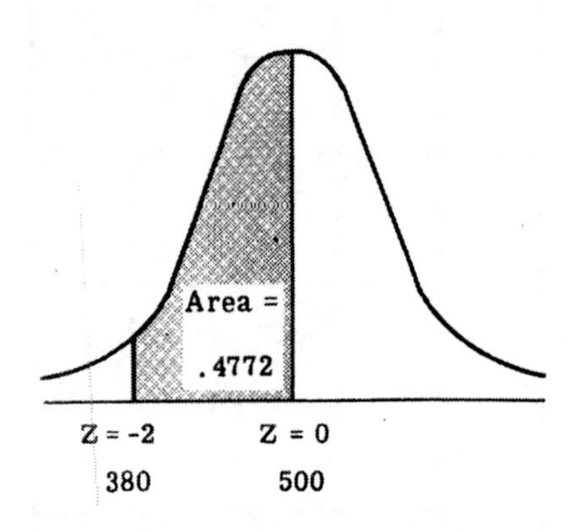

Step 4. The percentage of lights which have a lifetime greater than 380 is equal to the total shaded area or .4772 + the area under the positive half of the curve.:

.4772 + .5000 = .9772

97.72% of the lights have a life greater than 380 hours.

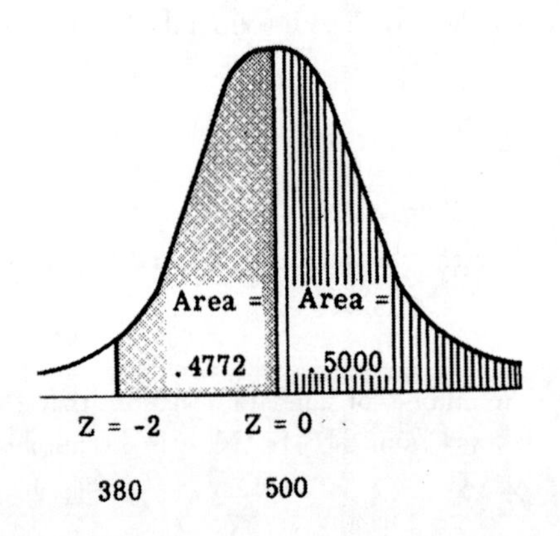

NORMAL DISTRIBUTION

$Z = \frac{X - \overline{X}}{\sigma}$	SECOND DECIMAL PLACE IN Z									
FIRST DECIMAL PLACE IN Z	.00	.01	.02	.03	.04	.05	.06	.07	.08	.09
.0	.0000	.0040	.0080	.0120	.0160	.0199	.0239	.0279	.0319	.0359
.1	.0398	.0438	.0478	.0517	.0557	.0596	.0636	.0675	.0714	.0753
.2	.0793	.0832	.0872	.0910	.0948	.0987	.1026	.1064	.1103	.1141
.3	.1179	.1217	.1255	.1293	.1332	. 1368	.1406	.1443	.1480	.1518
.4	.1554	.1591	.1628	.1664	.1700	.1736	.1772	.1808	.1844	.1879
.5	.1915	.1950	.1985	.2019	.2054	.2088	.2123	.2157	.2190	.2224
.6	.2257	.2291	.2324	.2357	.2389	.2422	.2454	.2486	.2517	.2549
.7	.2580	.2611	.2642	.2672	.2704	.2734	.2764	.2793	.2823	.2852
.8	.2882	.2910	.2939	.2967	.2995	.3023	.3051	.3078	.3106	.3133
.9	.3159	.3185	.3212	.3238	.3264	.3289	.3315	.3340	.3365	.3389
1.0	.3413	.3438	.3462	.3485	.3508	.3531	.3554	.3577	.3599	.3621
1.1	.3643	.3665	.3686	.3708	.3729	.3749	.3770	.3790	.3810	.3830
1.2	.3849	.3869	.3888	.3907	.3925	.3943	.3962	.3980	.3997	.4015
1.3	.4033	.4049	.4066	.4082	.4099	.4115	.4131	.4147	.4162	.4177
1.4	.4192	.4207	.4222	.4236	.4250	.4265	.4279	.4292	.4306	.4319
1.5	.4332	.4345	.4357	.4370	.4382	.4394	.4406	.4418	.4429	.4441
1.6	.4452	.4462	.4474	.4484	.4495	.4505	.4515	.4525	.4535	.4545
1.7	.4554	.4564	.4573	.4582	.4591	.4599	.4608	.4616	.4625	.4633
1.8	.4641	.4649	.4656	.4663	.4671	.4678	.4686	.4693	.4699	.4706
1.9	.4712	.4719	.4726	.4732	.4738	.4744	.4750	.4756	.4760	.4767
2.0	.4772	.4778	.4783	.4788	.4793	.4798	.4803	.4808	.4812	.4818
2.1	.4821	.4826	.4830	.4834	.4838	.4842	.4846	.4850	.4854	.4857
2.2	.4861	.4865	.4868	.4871	.4875	.4878	.4882	.4884	.4887	.4890
2.3	.4893	.4896	.4898	.4901	.4903	.4906	.4909	.4911	.4913	.4916
2.4	.4918	.4920	.4922	.4925	.4927	.4929	.4931	.4932	.4934	.4936
2.5	.4938	.4940	.4941	.4943	.4945	.4946	.4948	.4949	.4951	.4952
2.6	.4953	.4955	.4956	.4958	.4959	.4960	.4961	.4961	.4963	.4964
2.7	.4965	.4966	.4967	.4968	.4969	.4970	.4971	.4972	.4973	.4974
2.8	.4974	.4975	.4976	.4977	.4977	.4978	.4979	.4979	.4980	.4981
2.9	.4981	.4982	.4982	.4983	.4984	.4984	.4985	.4985	.4986	.4986
3.0	.4987	.4987	.4987	.4988	.4988	.4989	.4989	.4989	.4990	.4990
3.9	.5000	.5000	.5000	.5000	.5000	.5000	.5000	.5000	.5000	.5000

Solution (B):

The life, above which, we will find the best 20% of the headlights.?

Step 1. Draw a normal curve, label the MEAN and shade the desired area:

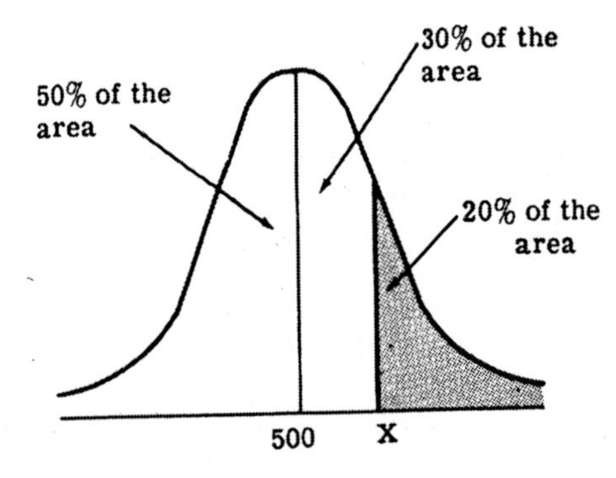

Step 2. We want the value of X above which 20% of the area of the curve lies. This means that 30% of the area lies between the centerline and X.

We are interested in the area from the centerline because the areas in the normal table are measured from the centerline to X.

Next, we want to find the value of Z that corresponds to 30% of the area under the curve.

We look in the body of the table for the area closest to 30% or .3000. The closest area is .2995 or 29.95% and its corresponding value of Z = .84. (Look up .8 row, .04 column in the table):

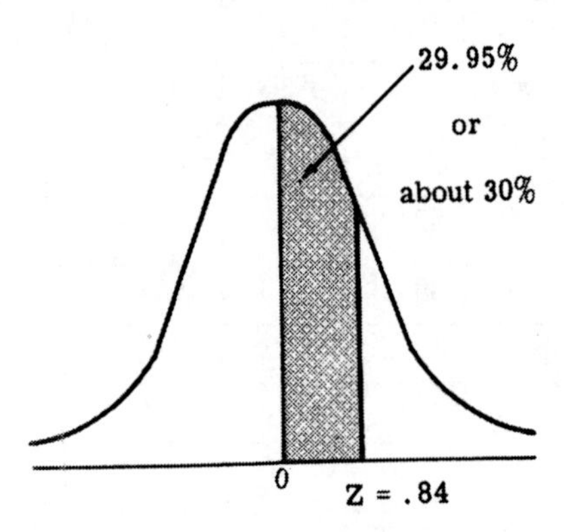

Step 3. The value of X that corresponds to Z = .84 standard deviations is found by substituting known value for Z, $\overline{X}$ and σ in the equation:

$$Z = \frac{X - \overline{X}}{\sigma}$$

$$Z = .84 = \frac{X - 500}{60}$$

$$(.84)\,(60) = X - 500$$

$$X = 500 + 50.4 = 550.4$$

Step 4. Therefore, since 30% of the area of the normal curve lies between the centerline and X = 550.4, 20% of the area must lie above this point. Thus, the life above which we will find the best 20% of the headlights is 550.4 hours:

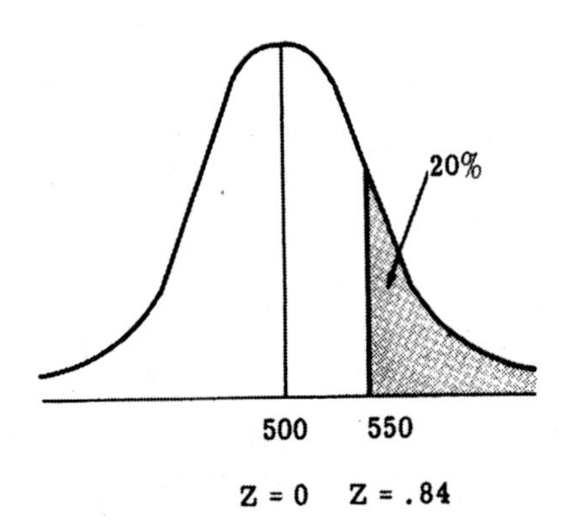

Solution (C):

Find the percentage of headlights which have a life between 440 and 530 hours:

Step 1. Draw a normal curve, label the MEAN, the values of X and shade the desired areas:

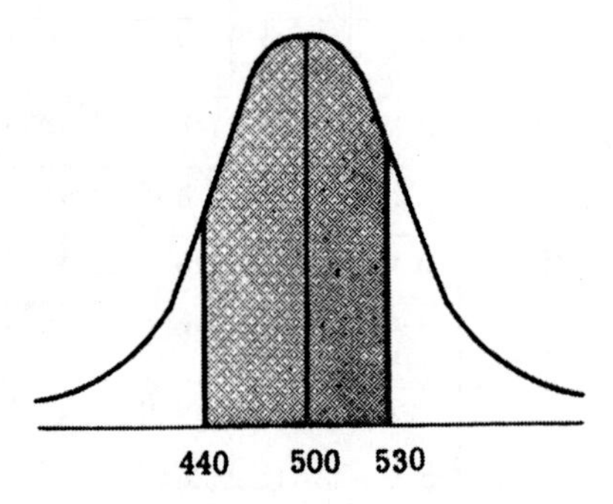

Step 2. Convert the values of X to their Z equivalents:

$$Z = \frac{X - \overline{X}}{\sigma} = \frac{440 - 500}{60} = \frac{-60}{60} = -1$$

$$Z = \frac{X - \overline{X}}{\sigma} = \frac{530 - 500}{60} = \frac{30}{60} = \frac{1}{2} = .5$$

Step 3. Look up the areas for the values Z = 1 and Z = .5. We cannot add the Z's together and then look up the areas; we must look up the individual Z's and then add the areas.

At Z = 1.00, area = .3413 (Z =1.00 in the 1.0 row, .00 column in the table).

At Z = .5, area = .1915 (Z = .50 in the .5 row, .00 column in the table).

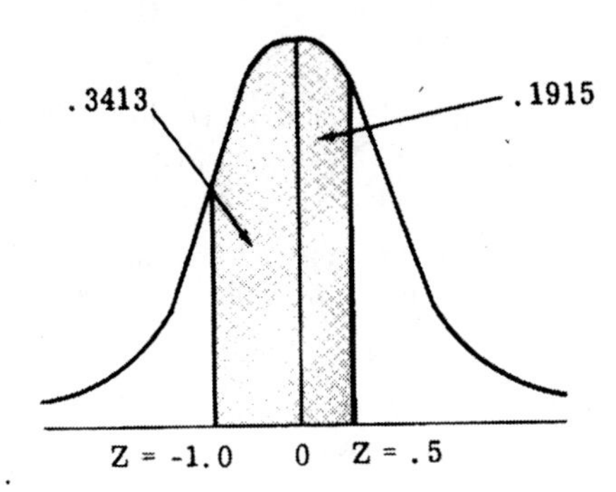

Step 4. The percentage of the headlights which have a life between 440 and 530 hours is equal to the area under the normal curve between these values:

Area between 440 and 500 hours = .3413

Area between 500 and 530 hours = .1915

Total area = .5328

Answer: 53.28% of the headlights.

1. SAMPLE NORMAL DISTRIBUTION PROBLEMS.

PROBLEM 1: The tensile strength of a certain plastic varies in accordance with a normal distribution. The MEAN = 1000 pounds and the standard deviation = 100 pounds. Find the probability that the strength of a sample piece of this plastic lies between 1100 and 1200 pounds:

Solution

Step 1. Draw a normal curve, label the MEAN, the values of X and shade the desired area:

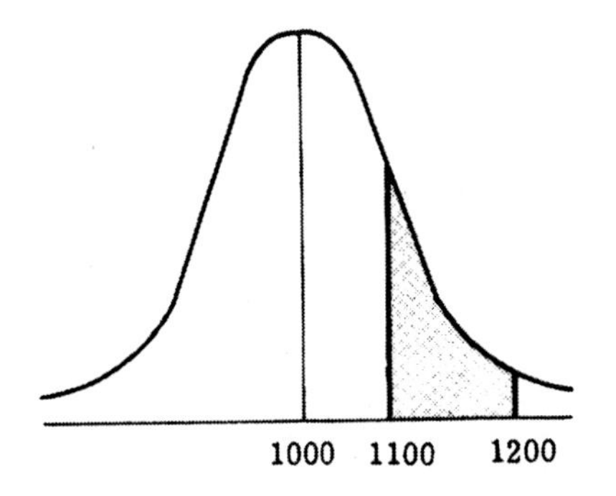

Step 2. Convert the value of X to their Z equivalents:

$$Z = \frac{X - \overline{X}}{\sigma} = \frac{1100 - 1000}{100}$$

$$= \frac{100}{100} = 1.0$$

$$Z = \frac{X - \overline{X}}{\sigma} = \frac{1200 - 1000}{100}$$

$$= \frac{200}{100} = 2$$

Step 3. Look up the areas for the Z values in the table of normal curve areas.

At Z = 1.0, (1.0 row, .00 column)
area = .3413

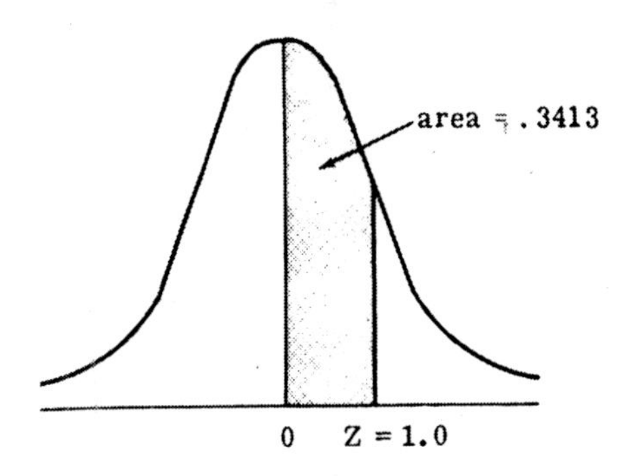

At Z = 2.0, area = .4772 (in table Z = 2.00 in the 2.0 row, .00 column).

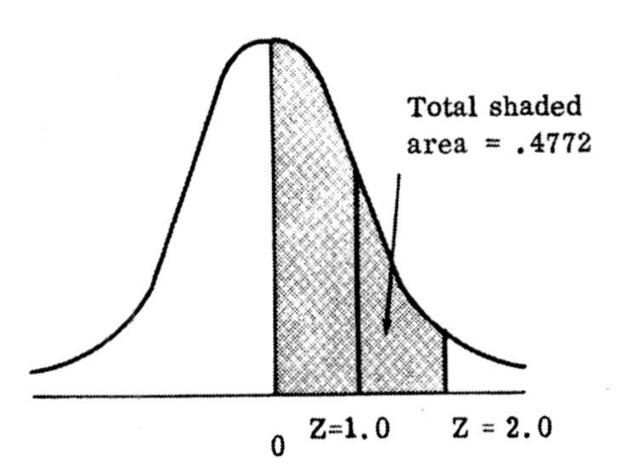

Step 4. Since both areas are measured from the centerline of the curve, the desired area between Z = 1.0 and Z = 2.0 is equal to (.4772 –.3413) = .1359

Therefore, the probability that the strength of a piece of plastic lies between 1100 and 1200 pounds is = .1359

PROBLEM 2: Find the normal curve area which lies to the right of

Z = –45

Solution

Step 1. Draw a normal curve, label Z, and shade the desired area:

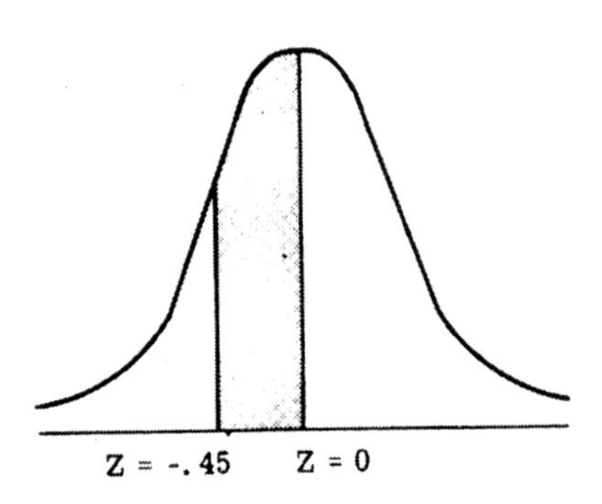

Step 2. From the table of normal curve areas, the area under the normal curve from the centerline to Z = –.45 is .1736

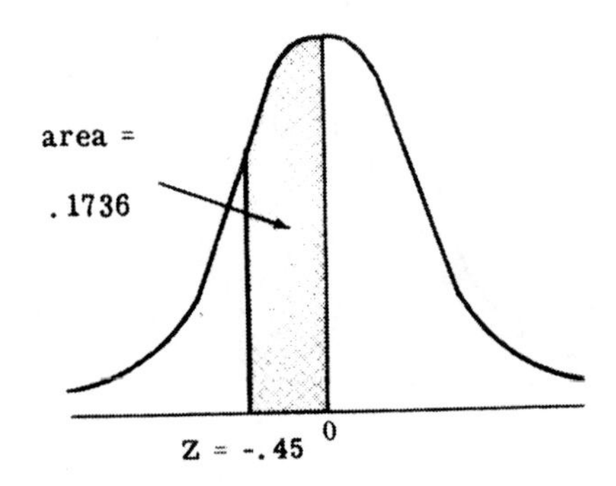

Step 3. The area to the right of Z = –.45 is equal to .1736 + the area of the other half of the normal curve. .1736 + .5000 = .6736. 67.36% of the area of a normal curve lies to the right of Z = –.45

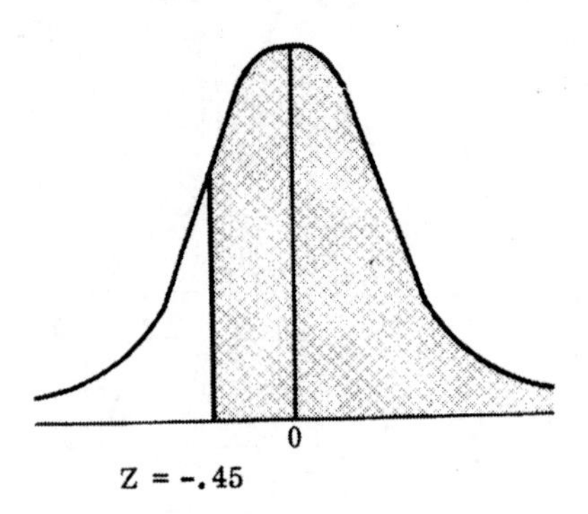

PROBLEM 3: The heights in inches of a large number of freshmen students had a MEAN of 70 inches and a standard deviation of 5 inches. Assuming a normal distribution, below which height will we find the shortest 15% of the class?

Solution

Step 1. Draw a normal curve, label the MEAN and shade the desired area:

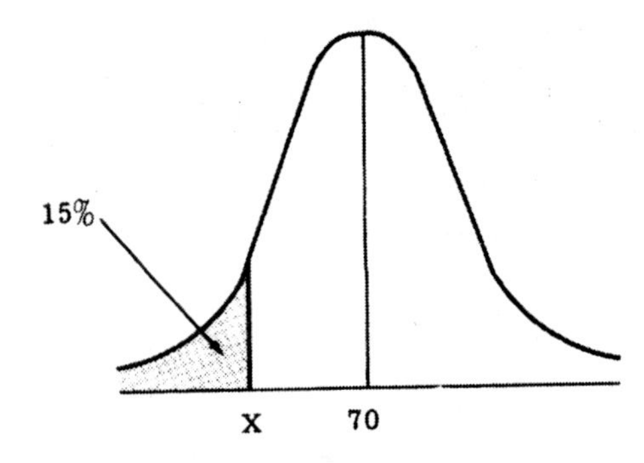

Step 2. We want the value of X below which 15% of the area of the curve lies. This means that 35% of the area lies between the centerline and X.

We are interested in the area from the centerline because the areas in the normal table are measured from the centerline to X.

Next we want to find the value of Z that corresponds to 35% of the area under the curve.

In the body of the table of areas, the area closest to 35% or .3500 is .3508 and its corresponding value of Z = –1.04. It's minus because Z is left of the centerline.

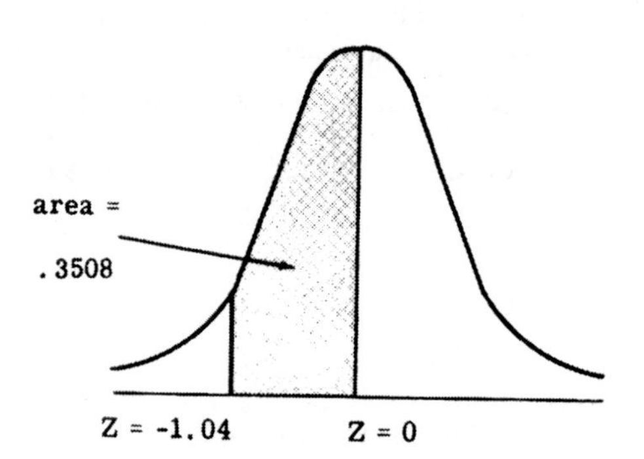

Step 3. The value of X that corresponds to Z = –1.04 standard deviations is found by substituting known values for Z, $\overline{X}$ and 0 in the equation:

$$Z = \frac{X - \overline{X}}{\sigma}$$

$$Z = -1.04 = \frac{X - 70}{5}$$

$$(-1.04)\,(5) = X - 70$$

$$X = 70 - 5.2$$

$$X = 64.8 \text{ inches}$$

Step 4. Since 35% of the area of the normal curve lies between the centerline and X = 64.8, 15% of the area must lie below this point.

Therefore, the shortest 15% of the freshmen students are under 64.8 inches in height.

PROBLEM 4: Find Z if the normal curve area to the left of Z = .9251:

Solution

Step 1. Draw the normal curve and shade in the desired area.

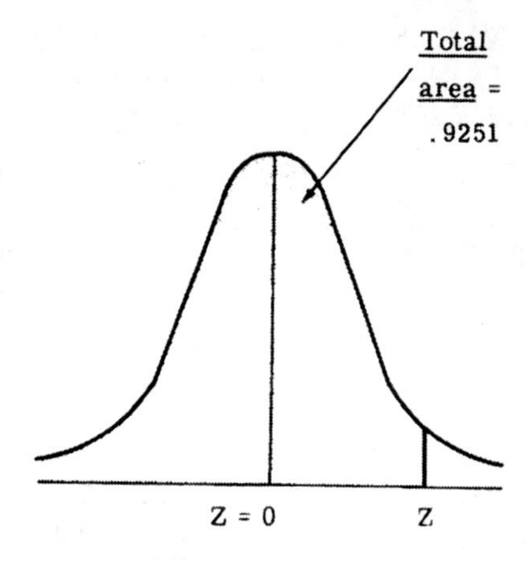

Step 2. The area to the left of the centerline is equal to .5000. Therefore the area from the centerline to Z is equal to:

$$.9251 - .5000 = .4251$$

Step 3. Find the value of Z corresponding to the area .4251 in the table of normal curve areas:

At .4251, Z = 1.44

11 POISSON DISTRIBUTION

HOW TO USE THE POISSON DISTRIBUTION TO COMPUTE PROBABILITIES

INTRODUCTION TO THE POISSON DISTRIBUTION. In the binomial probability distribution p represents the percentage of times that an event occurs, and q represents the percentage of times that an event does not occur. Thus, in order to determine p and q, we need to know the total number of occurrences, the total number of non-occurrences, and the total number of possible outcomes consisting of occurrences and non-occurrences.

There are problems where it is impossible to determine the number of times that an event does not occur. For example, in an electrical storm we can count the number of times that lightning flashes, but we do not know the number of times that lightning did not flash, or we can count the number of defects in a length of telephone cables but we do not know how many defects did not occur in the length, etc. We cannot use the binomial to determine probabilities for these examples, because we do not know the values for q and for the total number of possible outcomes.

We can, however, use a distribution called the Poisson distribution to determine the probability that an event will occur when the total number of possible outcomes is unknown.

1. THE GENERAL POISSON FORMULA: The general formula for computing probabilities by using the Poisson distribution is:

$$P_X = \frac{e^{-\lambda}\ \lambda^X}{X!} \text{ where:}$$

P = probability of the occurrence

P_X = probability of the occurrence of event X

X = event we want to occur (5 deaths out of 5,000 people injected with serum)

e = constant 2.718 =

$$\frac{1}{0!}+\frac{1}{1!}+\frac{1}{2!}+\frac{1}{3!}+\frac{1}{4!}+\frac{1}{5!}$$

! = factorial sign = product of all the consecutive integers from 1 to the number preceding the factorial sign:

Thus, 3! = 3 ! × 2 × 1 = 6

λ = Greek letter lambda = the average number of occurrences in the sample = NP = mean
Where:

N = sample size = number of trials

P = probability that an event will occur

2. TABLE OF VALUES OF $e^{-\lambda}$ FOR USE WITH THE POISSON DISTRIBUTION. The following is a three decimal place table of selected values of $e^{-\lambda}$:

λ	$e^{-\lambda}$		λ	$e^{-\lambda}$
0	1.000		1	.368
.1	.905		2	.135
.2	.819		3	.050
.3	.741		4	.018
.4	.670		5	.007
.5	.607		6	.002
.6	.549		7	.0009
.7	.497		8	.0003
.8	.449		9	.0001
.9	.407		10	.00004

3. SAMPLE POISSON PROBLEMS.

PROBLEM 1: If the probability that an individual will die from a certain disease is .005, what is the probability that out of 1,000 people having this disease 4 will die?

Solution

Step 1. $P_X = \frac{e^{-\lambda}\ \lambda^X}{X!}$

e = 2.718

X = 4 people

λ = (.005) (1000) = 5

Step 2. $P = \frac{2.718^{-5}\ (5)^4}{4!}$

$P = \frac{(.007)\ (625)}{4\times3\times2\times1} = \frac{4.4}{24}$

P = .183

PROBLEM 2: If an area has an average rainfall of 2 inches per month, what is the probability that it will get less than 3 inches in one month?

Solution

Less than 3 inches means the sum of probabilities of getting 0 inches of rain + 1 inch of rain + 2 inches of rain. These rainfalls are all less than three inches.

Step 1. Compute the probability of getting zero inches of rain:

$$P_X = \frac{e^{-\lambda} \ \lambda^X}{X!}$$

e = 2.718

X = 0 inches

λ = 2 inches/month

$$P_0 = \frac{(2.718)^{-2} \ (2)^0}{0!}$$

by definition: 0! = 1

$(2)^0$ = 1, because any number to the zero power is equal to 1

$$P = \frac{(.135) \ (1)}{(1)} = .135, \text{ probability of 0 inches of rain}$$

Step 2. Compute the probability of getting one inch of rain:

$$P_X = \frac{e^{-\lambda} \ \lambda^X}{X!}$$

e = 2.718

X = 1 inch

λ = 2 inches of rain/month

$$P_1 = \frac{(2.718)^{-2} \ (2)^1}{1!}$$

$$P_1 = \frac{(.135) \ (2)}{1} = .270, \text{ probability of 1 inch of rain}$$

Step 3. Compute the probability of getting two inches of rain:

$$P_X = \frac{e^{-\lambda} \ \lambda^X}{X!}$$

e = 2.718

X = 2 inches

λ = 2 inches of rain/month

$$P_2 = \frac{(2.718)^{-2} \ (2)^2}{2!}$$

$$P_2 = \frac{(.135) \ (4)}{2 \times 1} = \frac{.540}{2}$$

P_2 = .270, probability of 2 inches of rain.

Step 4. The probability of getting less than 3 inches of rain in a month is equal to the sum of the probabilities of getting 0, 1, and 2 inches of rain:

$$P = P_0 + P_1 + P_2$$

$$P = .135 + .270 + .270$$

P = .675, probability of less than 3 inches of rain

PROBLEM 3: If 5% of the products produced by a company are defective, what is the probability that there will be exactly 2 defective products in a shipment of 100?

Solution

Step 1. $$P_X = \frac{e^{-\lambda} \ \lambda^X}{X!}$$

e = 2.718

X = 2 defectives

λ = NP

λ = (100) (.05) = 5

Step 2. $$P = \frac{2.718^{-5} \ (5)^2}{2!}$$

$$P = \frac{(.007) \ (25)}{2 \times 1} = \frac{.175}{2}$$

P = .0875

PROBLEM 4: If a Pacific Island experiences an average of two earth tremors a month, what is the probability of getting more than two tremors in a month?

Solution

Step 1. More than two tremors also means not zero, not one and not two tremors in a month. The sum of the probabilities of all the number of tremors that can occur in a month is equal to 1. Therefore, the probability of getting more than two tremors is equal to one minus the sum of the probabilities of getting 0, 1, and 2 tremors.

Step 2. Probability of getting zero tremors:

$$P_X = \frac{e^{-\lambda} \ \lambda^X}{X!}$$

e = 2.718

X = 0 tremors

λ = 2 tremors a month

$$P = \frac{2.718^{-2} \ (2)^0}{0!}$$

$$P_X = \frac{(.135) \ (1)}{(1)} = .135$$

Step 3. Probability of getting <u>one</u> tremor?

$$P_X = \frac{e^{-\lambda}\ \lambda^X}{X!}$$

e = 2.718

X = 1 tremor

λ = 2 tremors a month

$$P = \frac{(2.718)^{-2}\ (2)^1}{1!}$$

$$P = \frac{(.135)\ (2)}{1} = .270$$

Step 4. Probability of getting two tremors?

$$P_X = \frac{e^{-\lambda}\ \lambda^X}{X!}$$

e = 2.718

X = 2 tremors

$\lambda = 2$

$$P = \frac{(2.718)^{-2}\ (2)^2}{2!}$$

$$P = \frac{(.135)\ (4)}{2 \times 1} = .270$$

Step 5. Probability of getting <u>more</u> than two tremors is equal to one minus the sum of the probability of getting 0, 1 and 2 tremors?

$$P = 1 - (P_0 + P_1 + P_2)$$

$$P = 1 - (.135 + .270 + .270)$$

$$P = 1 - .675 = .325$$

PROBLEM 5: In a traffic survey, 600 cars pass a certain point in one hour. What is the probability that 5 cars will pass the point in any one minute?

Solution

Step l. $$P_X = \frac{e^{-\lambda}\ \lambda^X}{X!}$$

e = 2.718

X = 5 cars

$\lambda = 600 \div 60 = 10$ cars/ a minute.

Step 2. $$P = \frac{2.718^{-10}\ (10)^5}{5!}$$

$$P = \frac{(.00005)\ (100000)}{5 \times 4 \times 3 \times 2 \times 1}$$

$$P = \frac{4}{120} = .038$$

PROBLEM 6: If a telephone switchboard receives an average of 120 incoming calls per hour, what is the probability of receiving at least 3 calls in any 5 minute period?

Solution

Step 1. <u>At least</u> 3 calls means the sum of the probabilities of receiving 3 or more calls. Since the sum of all the probabilities is one, it also means one <u>minus</u> the sum of the probabilities of receiving 0, 1 and 2 calls in the 5-minute period.

Step 2. Probability of receiving <u>zero</u> calls?

$$P_X = \frac{e^{-\lambda}\ \lambda^X}{X!}$$

e = 2.718

X = 0 calls

$\lambda = (120 \div 60) = 20$ calls/minute, (2 calls/ minute x 5 minutes) = 10 calls in a 5 minute period

$$P = \frac{2.718^{-10}\ (10)^0}{0!}$$

$$P = \frac{(.00004)\ (1)}{1} = .00004$$

Step 3. Probability of receiving <u>one</u> call?

$$P_X = \frac{e^{-\lambda}\ \lambda^X}{X!}$$

e = 2.718

X = 1 call

λ = 10 calls in a 5 minute period

$$P = \frac{2.718^{-10}\ (10)^1}{1!}$$

$$P = \frac{(.00004)\ (10)}{1} = .0004$$

Step 4. Probability of receiving two calls?

$$P_X = \frac{e^{-\lambda} \quad \lambda^X}{X!}$$

e = 2.718

X = 2 calls

λ = 10 calls/ 5 minutes

$$P = \frac{2.718^{-10} \quad (10)^2}{2.1}$$

$$P = \frac{(.00004) \quad (100)}{2 \times 1} = .002$$

Step 5. Probability of getting at least 3 calls is equal to the probability of not getting 0 calls + probability of not getting 1 call + probability of not getting 2 calls. In other words, it is equal to one minus the sum of the probabilities of getting 0, 1, and 2 calls:

$$P = 1 - (P_0 + P_1 + P_2)$$

$$P = 1 - (.00004 + .0004 + .002)$$

$$P = 1 - (.00244)$$

$$P = .99756$$

12 SAMPLING THEORY

HOW TO FIND CONFIDENCE INTERVALS FOR LARGE AND SMALL SAMPLES

INTRODUCTION TO SAMPLING. Sampling is basically the process of obtaining information about a large group of items by examining only a small number of them. For example, we can determine the average life of a brand of electric light bulbs by lighting some of them, until their elements burn out. Obviously, if we light all the bulbs, it would be very expensive and there would be no bulbs left to sell, because this is a destructive test.

POPULATION. Samples are taken from populations. Populations are a collection of items defined by a common characteristic. Some examples of populations are: all the parts produced by a machine in one day, all the doctors living in New York City, a truckload of oranges, a shipment of ball bearings, etc.

RANDOM SAMPLING. A sample is called a random sample when each item in the population has the same chance of being included in the sample.

1. STANDARD ERROR OF THE MEAN. Samples are usually taken from a population in order to get information about the mean and standard deviation of the population.

   The average means of samples taken from the same population will probably have different values from each other. If a large number of samples, whose size is greater than 30, is taken at random from a large population that has a normal distribution, then the values of the means of these samples will also be normally distributed. The average of all the sample means will be equal to the population mean. The standard deviation of the sample means is called the standard error of the mean.

   The standard error of the mean is equal to:

   $$S.E. = \frac{\sigma}{\sqrt{N}}$$

   σ = standard deviation of the population

   N = sample size

   We can see from this formula that as the sample size N gets larger, the standard error of the mean gets smaller. Thus, the larger the sample size, the more closely the sample will represent the population from which it was taken.

2. CONFIDENCE INTERVALS FOR LARGE SAMPLES (N > 30). In a normal distribution, about 68% of the values in the data are within ±1 standard deviation from the mean. About 95% of the values are within ±2 standard deviations from the mean, and about 99% of the values are within ±3 standard deviations from the mean.

   If the population has a normal distribution, then the values of the means of the samples taken from the population will also be normally distributed. Since the average of all the sample means is equal to the population mean, we can expect that the values of 68% of all the sample means will be within ±1 standard deviation from the true population mean. Thus, the 68% confidence interval for large samples is equal to $\overline{X}$ ±1 S.E., where $\overline{X}$ = population mean, and S.E. = standard error of the mean.

   The values of about 95% of all the sample means will be within ±2 standard deviations from the true population mean. Thus, the 95% confidence interval for large samples is: $\overline{X}$ ±2 S.E. Similarly, the 99% confidence interval for large samples is $\overline{X}$ ±3 S.E.

3. CONFIDENCE INTERVALS FOR SMALL SAMPLES (N < 30). The values of the means of large samples, whose size is greater than 30, will be normally distributed. The values of the means of small samples, whose size is less than 30, will not be normally distributed. They will be distributed in accordance with the T or students' distribution.

   The T distribution resembles the bell shape of the normal distribution. There is actually a different distribution for each sample size. The T table contains a tabulation of the proportion of the areas under the T distribution at various confidence levels. For example, the .05 or 5% confidence level means that only 5% of the means of the samples are more than the T Table value of standard deviations away from the population mean. Conversely, 95% of the means will be within the T Table value of standard deviations from the population mean. For the T Table the sample sizes are converted to degrees of freedom by subtracting one from them. Thus, a sample of size N = 10 has (N – 1) or 9 degrees of freedom.

   The 95% confidence interval for small samples is equal to $\overline{X}$ ± (T Table value) S.E. The T Table values are found in the .05 column and the (N – 1) row of the T Table for this interval. The following chapter, "DIFFERENCE BETWEEN TWO MEANS," contains a "T Table."

Using the T Table for a sample size =10, the degrees of freedom are (10 – 1) = 9, and we find that at the .05 confidence level, 5% of the sample means will be more than 2.26 standard deviations away from the population mean.

4. <u>SAMPLE CONFIDENCE INTERVAL PROBLEMS</u>.

PROBLEM 1: Find the 95% confidence intervals for the following data:

$$\overline{X} = 50, \sigma = 4.5 \text{ and } N = 81$$

Solution

Step 1: Sample size is greater than 30, so it is a large sample. The 95% confidence interval is:

$$\overline{X} \pm 2 \text{ S. E.}$$

$$\text{S.E.} = \frac{\sigma}{\sqrt{N}} = \frac{4.5}{\sqrt{81}} = \frac{4.5}{9} = .5$$

Step 2. Find $\overline{X} \pm 2$ S.E.

$$\overline{X} + 2 \text{ S.E.} = 50 + 2\,(.5) = 51$$

$$\overline{X} - 2 \text{ S.E.} = 50 - 2\,(.5) = 49$$

The 95% confidence interval is 49 to 51. This means that we are 95% sure <u>or</u> confident that this interval <u>includes</u> the population mean.

PROBLEM 2: Find the 99% confidence intervals for the following data:

$$\overline{X} = 20, \sigma = 5, N = 25$$

Solution

Step 1. Sample size is less than 30, so it is a <u>small</u> sample. The 99% confidence interval involves the T Table at the .01 or 1% interval since (100% – 99%) = 1%:

$$\text{S.E.} = \frac{\sigma}{\sqrt{N}} = \frac{5}{\sqrt{25}} = \frac{5}{5} = 1$$

Step 2. Find degrees of freedom.

Degrees of freedom = (N – 1)

= (25 – 1) = 24

Step 3. In the T Table at the .01 level and degrees of freedom = 24, the T value is equal to 2.80 standard deviations.

The 99% confidence intervals are therefore equal to $\overline{X} \pm 2.80$ S.E.

$$\overline{X} + 2.80 \text{ S.E.} = 20 + 2.80\,(1) = 22.80$$

$$\overline{X} - 2.80 \text{ S.E.} = 20 - 2.80\,(1) = 17.20$$

The 99% confidence interval is 17.20 to 22.80

PROBLEM 3: Find the 99% confidence intervals for the following data:

$$\overline{X} = 156, \sigma = 20, \text{ and } N = 100$$

Solution

Step 1. Sample size is greater than 30, so it is a large sample. The 99% confidence interval is $\overline{X} \pm 3$ S.E.

$$\text{S.E.} = \frac{\sigma}{\sqrt{N}} = \frac{20}{\sqrt{100}} = \frac{20}{10} = 2$$

Step 2. Find $\overline{X} \pm 3$ S.E.

$$\overline{X} + 3 \text{ S.E.} = 156 + 3(2) = 162$$

$$\overline{X} - 3 \text{ S.E.} = 156 - 3(2) = 150$$

The 99% confidence interval is 150 to 162

13 DIFFERENCE BETWEEN TWO MEANS

HOW TO USE THE "T" TEST TO ANALYZE THE DIFFERENCE BETWEEN TWO SAMPLE MEANS

INTRODUCTION TO THE ANALYSIS OF THE DIFFERENCE BETWEEN SAMPLE MEANS. It can be expected that any two samples, taken at random from the same population, will probably have different average means. The difference between them can be due simply to chance, or it can indicate that some change has taken place in the population during the sampling interval.

Thus, by analyzing the difference between two sample means, it is possible to determine if the samples were taken from the same or from different populations.

INTRODUCTION TO THE NORMAL DEVIATE. Two samples taken from the same population are called a sample pair, and the difference between their means is equal to:

$$\left(\overline{X}_1 - \overline{X}_2\right)$$

When a large number of sample pairs are drawn from a population, the differences between the means of the pair, the $\left(\overline{X}_1 - \overline{X}_2\right)$, are normally distributed, and the average of all the differences is equal to zero.

The standard deviation of the differences between the means of the sample pairs is called the Normal Deviate.

1. FORMULA FOR THE NORMAL DEVIATE: The formula for calculating the Normal Deviate or standard error of the difference between the means of a sample pair is:

$$\sigma_D = \sqrt{\frac{\sigma_1^2}{N_1} + \frac{\sigma_2^2}{N_2}}$$

σ_D = Normal Deviate

σ_D = Standard error of the difference between two sample means

σ_1 = Standard deviation of sample #1

σ_1^2 = Variance of sample #1

N_1 = Size of sample #1

σ^2 = Standard deviation of sample #2

σ_2^2 = Variance of sample #2

N_2 = Size of sample #2

INTRODUCTION TO THE "T" TABLE. The difference between two sample means $\left(\overline{X}_1 - \overline{X}_2\right)$ is compared with the Normal Deviate σ_D in order to determine if the difference is due to chance, or to the fact that the samples came from different populations.

This comparison is called the T RATIO and is equal to:

$$T = \frac{\overline{X}_1 - \overline{X}_2}{\sigma_D}$$

$$T = \frac{\overline{X}_1 - \overline{X}_2}{\sqrt{\frac{\sigma_1^2}{N_1} + \frac{\sigma_2^2}{N_2}}}$$

$\overline{X}_1$ = Mean of sample #1

$\overline{X}_2$ = Mean of sample #2

σ_D = Normal Deviate

or

σ_D = Standard error of the difference between two sample means

The computed T ratio is then compared to a table of standard T ratio values to determine if the differences are significant or just due to chance sampling errors.

The T Table contains critical values of the T ratio at different confidence or probability levels, for various degrees of freedom.

The .10 or 10% confidence level means that the T Table value will be exceeded in 10 times out of 100. The .05 or 5% level will be exceeded in 5 out of 100 times, etc.

It is assumed that the samples are drawn from different populations when the computed T ratio exceeds the T Table value at the 5% confidence level.

The degrees of freedom in the T test are equal to: (size of sample #1 minus 1) + (size of sample #2 minus 1) = $(N_1 - 1) + (N_2 - 1) = (N_1 + N_2 - 2)$.

The T ratio is sometimes called the students' T because its developer, Mr. W. S. Gosset, originally published his work under the name "student."

1. <u>THE T TABLE:</u> Shown to the right is a two-decimal place T table containing selected confidence or probability of occurrence levels.

2. <u>PROCEDURE FOR ANALYZING THE DIFFERENCE BETWEEN SAMPLE MEANS.</u> The method for using the T test to analyze the difference between sample means can best be illustrated through the solution of a sample problem.

EXAMPLE: In a toothpaste test, the 90 people in group A had an average of 10 cavities each, with a standard deviation of 3. The 80 people in group B averaged 8 cavities each with a standard deviation of 2. Is there a significant difference in the results of the two groups?

Solution

Step l. Arrange the data of the problem into a table:

	N	$\overline{X}$	σ	σ^2
Sample A:	90	10	3	9
Sample B:	80	8	2	4

Step 2. $$\text{T ratio} = \frac{\overline{X}_1 - \overline{X}_2}{\sigma_D}$$

$$\text{T ratio} = \frac{\overline{X}_1 - \overline{X}_2}{\sqrt{\frac{\sigma_1^2}{N_1} + \frac{\sigma_2^2}{N_2}}}$$

$$\text{T ratio} = \frac{10 - 8}{\sqrt{\frac{9}{90} + \frac{4}{80}}}$$

Degrees of Freedom	Confidence Levels		
	P =. 1 or 10%	P =.05 or 5%	P =.01 or 1%
1	6.31	12.71	63.66
2	2.92	4.30	9.93
3	2.35	3.18	5.84
4	2.13	2. 78	4.60
5	2.02	2.57	4.03
6	1.94	2.45	3.71
7	1.89	2.37	3.50
8	1.86	2. 31	3.36
9	1.83	2.26	3.25
10	1.81	2. 23	3.17
11	1.79	2. 20	3.10
12	1.78	2.18	3.06
13	1.77	2.16	3.01
14	1.76	2.15	2. 98
15	1.75	2.13	2. 95
16	1.74	2.12	2.92
17	1.74	2.11	2. 90
18	1.73	2.10	2.88
19	1.73	2.09	2.86
20	1.72	2.09	2. 84
21	1.72	2.08	2.83
22	1.72	2.07	2.82
23	1.71	2.07	2.81
24	1.71	2. 06	2. 80
25	1.71	2.06	2.79
	1.65	1.96	2.56

$$T = \frac{+2}{\sqrt{.1 + .05}}$$

$$T = \frac{+2}{\sqrt{.15}} = \frac{+2}{.39}$$

$$T = +5.13$$

Step 3. Calculate the degrees of freedom:

Degrees of freedom $= (N_1 + N_2 - 2)$

$= (90 + 80 - 2)$

$= 170 - 2 = 168$

Step 4. Compare the computed T ratio = 5.13 to the T table value at 168 or ∞ degrees of freedom:

From the table at the .05 level T = 1.96. Since the computed value is larger, the difference is assumed to be significant and not due to chance.

3. SAMPLE PROBLEMS OF THE ANALYSIS OF MEANS.

PROBLEM 1: On a statistics final, the 40 students in section A had an average grade of 75, with a standard deviation of 4. The 30 students in section B had an average grade of 80, with a standard deviation of 9. Is there a significant difference between the two sections?

Solution

Step 1. Arrange the data of the problem into a table:

	N	$\overline{X}$	σ	σ^2
Section A:	40	75	4	16
Section B:	30	80	9	81

Step 2. $\text{T ratio} = \dfrac{\overline{X}_1 - \overline{X}_2}{\sigma_D}$

$$T = \frac{\overline{X}_1 - \overline{X}_2}{\sqrt{\dfrac{\sigma_1^2}{N_1} + \dfrac{\sigma_2^2}{N_2}}}$$

$$T = \frac{40 - 30}{\sqrt{\dfrac{16}{40} + \dfrac{81}{30}}}$$

$$T = \frac{10}{\sqrt{.4 + 2.7}} = \frac{10}{\sqrt{3.1}}$$

$T = 5.7$

Step 3. Calculate the degrees of freedom:

Degrees of freedom $= (N_1 + N_2 - 2)$

$= (40 + 30 - 2)$

$= 68$

Step 4. Compare the computed T ratio of 5.7 to the T table value at 68 or ∞ degrees of freedom.

From the table at the .05 level, T = 1.96. Since the computed value is greater than the table value, the difference between the sections is significant.

PROBLEM 2: Samples are taken from two machines producing the same part. The parts from machine #1 have an average diameter of .50 inches, with a standard deviation of .10 inches. The parts from machine #2 have an average diameter of .52 inches, with a standard deviation of .15 inches. If a sample of 50 items were taken from the first machine and a sample of 60 items from the second, do the results obtained indicate a difference between the two machines?

Solution

Step 1. Arrange the data of the problem into a table:

	N	$\overline{X}$	σ	σ^2
Machine #1	50	.50	.10	.010
Machine #2	60	.52	.15	.023

Step 2. $T = \dfrac{\overline{X}_1 - \overline{X}_2}{\sigma_D}$

$$T = \frac{\overline{X}_1 - \overline{X}_2}{\sqrt{\dfrac{\sigma_1^2}{N_1} + \dfrac{\sigma_2^2}{N_2}}}$$

$$T = \frac{50 - 60}{\sqrt{\dfrac{.01}{50} + \dfrac{0.023}{60}}}$$

$$T = \frac{-10}{\sqrt{.0002 + .0004}}$$

$$T = \frac{-10}{\sqrt{.0006}} = \frac{-10}{.024} = -416.7$$

T = 416.6 (Ignore the sign.)

Step 3. Calculate the degrees of freedom:

Degrees of freedom $= (N_1 + N_2 - 2)$

$= (50 + 60 - 2)$

$= 108$

Step 4. Compare the computed T ratio of 416 to the table value at 108 or ∞ degrees of freedom:

From the table at the .05 level T = 1.96. Since the computed value is greater than the table value, the difference between the two machines is significant.

PROBLEM 3: Two samples are taken from a truckload of apples. The apples in sample #1 have an average weight of 20 ounces, and a standard deviation of 2 ounces. The apples in sample #2 have an average weight of 25 ounces, and a standard deviation of 3 ounces. Is there a significant different between the two samples if each sample contained 50 apples?

Solution

Step 1. Arrange the data of the problem into a table:

	N	$\overline{X}$	σ	σ^2
Sample #1	50	20	2	4
Sample #2	50	25	3	9

Step 2. $$T = \frac{\overline{X}_1 - \overline{X}_2}{\sqrt{\frac{\sigma_1^2}{N_1} + \frac{\sigma_2^2}{N_2}}}$$

$$T = \frac{20 - 25}{\sqrt{\frac{4}{50} + \frac{9}{50}}} = \frac{-5}{\sqrt{.08 + .18}}$$

$$T = \frac{-5}{\sqrt{2.6}} = \frac{-5}{.51} = -9.8$$

T = 9.8 (Ignore the sign.)

Step 3. Calculate the degrees of freedom;

Degrees of freedom $= (N_1 + N_2 - 2)$

$= (50 + 50 - 2)$

$= 98$

Step 4. Compare the computed T ratio of 9.8 to the table value at 98 or ∞ degrees of freedom:

From the table at the .05 level T = 1.96. Since the computed value is more than the table value, the difference between the samples is significant.

PROBLEM 4: An automobile manufacturer suspects that the night-shift workers are applying a thicker coat of paint to the cars than the day-shift workers. To keep costs down, the manufacturer would like to have the paint applied as uniformly as possible. An investigation of the paint thickness disclosed the following:

Shift	Number of Measurement	Main Paint Thickness	Standard Deviation
	N	$\overline{X}$	σ
Day	200	. 002 in.	.0005
Night	250	. 003 in.	.0005

Solution

Step 1. Arrange the data of the problem into a table:

	N	$\overline{X}$	σ	σ^2
Day Shift	200	.002	.0005	.00000025
Night Shift	250	.003	.0005	.00000025

Step 2. $$T = \frac{\overline{X}_1 - \overline{X}_2}{\sqrt{\frac{\sigma_1^2}{N_1} + \frac{\sigma_2^2}{N_2}}}$$

$$T = \frac{.002 - .003}{\sqrt{\frac{.00000025}{200} + \frac{.00000025}{250}}}$$

$$T = \frac{-.001}{\sqrt{.000000002}} = \frac{-.001}{.00005} = -20$$

T = 20 (Ignore sign.)

Step 3. Calculate the degrees of freedom:

Degrees of freedom $= (N_1 + N_2 - 2)$

$= (200 + 250 - 2)$

$= 448$

Step 4. Compare the computed T ratio of 20 to the table value at 448 or ∞ degrees of freedom:

From the table at the .05 level, T = 1.96. Since the computed value is more than the table value, the difference is significant.

PROBLEM 5: Over a six-month period Plant A had an average of 30 accidents per 100 employees, with a standard deviation of five accidents. Plant B had an average of 50 accidents per 100 employees, with a standard deviation of two accidents. Is there a significant difference in the accident rate of the two plants?

Solution

Step 1. Arrange the data of the problem into a table:

	N	$\overline{X}$	σ	σ^2
Plant A	100	30	5	25
Plant B	100	50	2	4

Step 2.

$$T = \frac{\overline{X}_1 - \overline{X}_2}{\sqrt{\frac{\sigma_1^{\,2}}{N_1} + \frac{\sigma_2^{\,2}}{N_2}}}$$

$$T = \frac{30 - 50}{\sqrt{\frac{25}{100} + \frac{4}{100}}}$$

$$T = \frac{-20}{\sqrt{.25 + .04}} + \frac{-20}{\sqrt{.29}}$$

$$T = \frac{-20}{.54} = -37$$

T = 37 (Ignore the sign.)

Step 3. Calculate the degrees of freedom:

Degrees of freedom $= (N_1 + N_2 - 2)$

$= (100 + 100 - 2)$

$= 198$

Step 4. Compare the computed T ratio = 37 to the T table value of 198 or ∞ degrees of freedom:

From the table at the .05 level, T = 1.96. Since the computed value of T is larger, the difference is significant.

14 ANALYSIS OF VARIANCE

HOW TO DETERMINE IF THERE IS A SIGNIFICANT DIFFERENCE BETWEEN MORE THAN TWO AVERAGE MEANS

INTRODUCTION TO THE ANALYSIS OF VARIANCE. The variance you will remember is equal to σ^2. The analysis of variance is a technique for determining if there is a significant difference between two or more average means.

This difference is evaluated by comparing the variation within the samples to the variation between these Samples.

The measure of the variation within the samples is called the within sample variance. It is an over-all variance computed for all the data being compared.

The measure of the variation between the Samples is called the between sample variance. It is the variance obtained when only the average means being compared are used as the data.

1. FORMULAS USED IN THE ANALYSIS OF VARIANCE.

Formula A: The formula for calculating the within sample variance is:

$$\sigma_X^2 = \frac{\Sigma(X - \overline{X})^2}{\Sigma(N - 1)} \quad \text{where:}$$

$\overline{X}$ = Mean of the Sample

$(X - \overline{X})$ = Deviation or difference between a number in the Sample and its corresponding mean

$(X - \overline{X})^2$ = Deviations squared

$\Sigma(X - \overline{X})^2$ = Sum of all the squared deviations of all the samples

N = Number of items in the Sample = the Sample size

N – 1 = Number of degrees of freedom in the sample = Sample size minus one

Σ(N – 1) = Sum of all the degrees of freedom of all the Samples

Formula B: The formula for calculating the between Sample variance is:

$$\sigma_X^2 = \frac{(X - \overline{X})^2}{(N - 1)} \quad \text{where:}$$

$\overline{X}$ = Mean of the Sample

$\overline{\overline{X}}$ = Sum of the Sample means divided by the number of Samples = population mean

$(\overline{X} - \overline{\overline{X}})$ = Deviation or difference between the Sample means $(\overline{X})$ and the population mean $(\overline{\overline{X}})$

$(\overline{X} - \overline{\overline{X}})^2$ = Deviations squared

$\Sigma(\overline{X} - \overline{\overline{X}})^2$ = Sum of all the squared deviations

N = Number of Samples

(N – 1) = Number of degrees of freedom between the Samples = number of Samples minus one

INTRODUCTION TO THE "F" TABLE. The between sample variance is compared to the within sample variance in the ratio:

$$\text{F ratio} = \frac{\text{"between" Sample variance}}{\text{"within" Sample variance}}$$

This ratio is then compared to a table of standard ratio values to determine if it is significant. The table of ratios is called the F table in honor of R. A. Fisher. It contains the critical ratio values of freedom.

1. F TEST TABLE FOR THE 1% CONFIDENCE LEVEL: The one-decimal place F table (see following page) contains ratio values at the 1% confidence or probability level. This means that a computed ratio would be larger than the F table value in less than one out of a hundred times. The ratios are shown for selected degrees of freedom for the "within" and the "between" sample variances.

PROCEDURE FOR COMPUTING THE ANALYSIS OF VARIANCE. The procedure for calculating the analysis of variance can best be illustrated through the solution of a Sample problem:

$F = \frac{\text{"between" var.}}{\text{"within" var.}}$		DEGREES OF FREEDOM — "between" Sample Variance										
		1	2	3	4	5	6	7	8	9	10	∞
Degrees of Freedom "within" Sample Variance	1	161	200	216	225	230	234	237	239	241	242	254
	2	18.5	19.0	19.2	19.2	19.3	19.3	19.4	19.4	19.4	19.4	19.5
	3	10.2	9.5	9.3	9.1	9.0	8.9	8.9	8.9	8.8	8.8	8.5
	4	7.7	6.9	6.6	6.4	6.3	6.2	6.1	6.0	6.0	6.0	5.6
	5	6.6	5.8	5.4	5.2	5.1	5.0	4.9	4.8	4.8	4.7	4.4
	6	6.0	5.1	4.8	4.5	4.4	4.3	4.2	4.2	4.1	4.1	3.7
	7	5.6	4.7	4.4	4.1	4.0	3.9	3.8	3.7	3.7	3.6	3.2
THE "F" TABLE	8	5.3	4.5	4.1	3.8	3.7	3.6	3.5	3.4	3.4	3.4	2.9
	9	5.1	4.3	3.9	3.6	3.5	3.4	3.3	3.2	3.2	3.1	2.7
	10	5.0	4.1	3.7	3.5	3.3	3.2	3.1	3.1	3.0	3.0	2.5
	∞	3.8	3.0	2.6	2.4	2.2	2.1	2.0	1.9	1.9	1.8	1.0

EXAMPLE: Determine if the average means of the following four samples are significantly different:

Sample A	Sample B	Sample C	Sample D
9	4	2	4
7	5	1	4
2	6	2	4
	5	7	

Solution

Step 1. Calculate the "within" Sample variance:

$$\sigma_X^2 = \frac{\text{Sum of the squared deviations from } \overline{X}}{\text{Total number of degrees of freedom}}$$

$$\sigma_X^2 = \frac{\Sigma(X - \overline{X})^2}{\Sigma(N - 1)}$$

Needed: $\overline{X}$, $X - \overline{X}$, $(X - \overline{X})^2$, and $\Sigma(X - \overline{X})^2$

for each Sample, the degrees of freedom for each Sample and the total degrees of freedom.

(a) Find the average means for each of the Samples;

Sample A: $\overline{X} = \frac{\Sigma X}{N} = \frac{9+7+2}{3} = \frac{18}{3} = 6$

Sample B: $\overline{X} = \frac{\Sigma X}{N} = \frac{4+5+6+5}{4} = \frac{20}{4} = 5$

Sample C: $\overline{X} = \frac{\Sigma X}{N} = \frac{2+1+2+7}{4} = \frac{12}{4} = 3$

Sample D: $\overline{X} = \frac{\Sigma X}{N} = \frac{4+4+4}{3} = \frac{12}{3} = 4$

(b) Arrange the data into a table, compute the deviations from $\overline{X}$ for each sample and then square the deviations. (See this and the following page);

(c) Calculate the degrees of freedom for each Sample;

Degrees of freedom = (Sample size minus one) = N − 1;

Sample A: $N_1 - 1 = (3 - 1) = 2$

Sample B: $N_2 - 1 = (4 - 1) = 3$

Sample A		
$\overline{X} = 6$		
X	$X - \overline{X}$	$(X - \overline{X})^2$
9	+3	9
7	+1	1
2	−4	16
$\Sigma(X - \overline{X})^2 = 26$		

Sample B		
$\overline{X} = 5$		
X	$X - \overline{X}$	$(X - \overline{X})^2$
4	−1	1
5	0	0
6	+1	1
5	0	0
$\Sigma(X - \overline{X})^2 = 2$		

Sample C		
$\overline{X} = 3$		
X	$X - \overline{X}$	$(X - \overline{X})^2$
2	–1	1
1	–2	4
2	–1	1
7	+4	16
$\Sigma(X - \overline{X})^2 = 22$		

Sample D		
$\overline{X} = 4$		
X	$X - \overline{X}$	$(X - \overline{X})^2$
4	0	0
4	0	0
4	0	0
$\Sigma(X - \overline{X})^2 = 0$		

Sample C: $N_3 - 1 = (4 - 1) = 3$

Sample D: $N_4 - 1 = (3 - 1) = 2$

Total degrees of freedom =

$\Sigma (N - 1) = (2 + 3 + 3 + 2) = 10$

(d) Compute the "within" sample variance:

$$\sigma_X^2 = \frac{\text{Sum of all the squared deviations from } \overline{X}}{\text{Total number of degrees of freedom}}$$

$$\sigma_X^2 = \frac{\Sigma(X_1 - \overline{X})^2 + \Sigma(X_2 - \overline{X})^2 + \Sigma(X_3 - \overline{X})^2 + \Sigma(X_4 - \overline{X})^2}{(N_1 - 1) + (N_2 - 1) + (N_3 - 1) + (N_4 - 1)}$$

$$\sigma_X^2 = \frac{26 + 2 + 22 + 0}{2 + 3 + 3 + 2} = \frac{50}{10} = 5$$

Step 2. Calculate the "between" mean variance:

$$\sigma_{\overline{X}}^2 = \frac{\text{Sum of the squared deviations from } \overline{X}}{\text{Number of degrees of freedom}}$$

$$\sigma_{\overline{X}}^2 = \frac{\Sigma(X - \overline{X})^2}{(N - 1)}$$

Needed:

$\overline{\overline{X}}$ (average of the means of the Samples), $\overline{X} - \overline{\overline{X}}$, $\left(\overline{X} - \overline{\overline{X}}\right)^2$,

$\Sigma\left(\overline{X} - \overline{\overline{X}}\right)^2$, and the number of degrees of freedom

(a) Calculate the average of the means of the Samples:

Sample A: $\overline{X}_1 = 6$

Sample B: $\overline{X}_2 = 5$

Sample C: $\overline{X}_3 = 3$

Sample D: $\overline{X}_4 = 4$

$$\overline{\overline{X}} = \frac{\Sigma\overline{X}}{N} = \frac{6 + 5 + 3 + 4}{4} = \frac{18}{4} = 4.5$$

(b) Arrange the data into a table. Compute the deviations from $\overline{\overline{X}}$ for each sample mean and then square the deviations.

Average of the Sample Means = $\overline{\overline{X}} = 4.5$		
Sample mean	Deviation	Deviation squared
$(\overline{X})$	$(\overline{X} - \overline{\overline{X}})$	$(\overline{X} - \overline{\overline{X}})^2$
6	+1.5	2.25
5	+ .5	.25
3	−1.5	2.25
4	− .5	.25
$\Sigma(\overline{X} - \overline{\overline{X}})^2$	=	5.00

(c) Calculate the degrees of freedom:

Degrees of freedom = (number of samples minus one) = (4 – 1) = 3

(d) Compute the "between" Sample variance.

$$\sigma_{\overline{X}}^2 = \frac{\text{Sum of the squared deviations from } \overline{\overline{X}}}{\text{Number of degrees of freedom}}$$

$$\sigma_{\overline{X}}^2 = \frac{5.00}{3} = 1.67$$

Step 3. Compare the within sample variance to the between sample variance by computing the ratio between them:

$$F = \frac{\text{between Sample variance}}{\text{within Sample variance}} = \frac{1.67}{5} = .334$$

Step 4. Compare the computed ratio of .334 to the F table value at the between sample degrees of freedom = 3, and the within Sample degrees of freedom = 10. (in table, column 3, row 10):

The table ratio = 3.7. Since the computed ratio is less than 3.7, there is no significant difference between the two means.

1. SAMPLE ANALYSIS OF VARIANCE PROBLEMS.

PROBLEM 1: Determine if the means of the following two Samples are significantly different:

Sample A	Sample B
20	30
80	40
50	50

Solution

Step 1. Calculate the "within" Sample variance:

$$\sigma_X^2 = \frac{\Sigma(X - \overline{X})^2}{\Sigma(N - 1)}$$

Needed the:

X, $X - \overline{X}$, $(X - \overline{X})$ and $\Sigma(X - \overline{X})^2$, for each Sample, the degrees of freedom for each sample and the total degrees of freedom.

(a) Find the average means for each of the Samples:

$$\text{Sample A: } \overline{X} = \frac{\Sigma X}{N} = \frac{20 + 80 + 50}{3} = \frac{150}{3} = 50$$

$$\text{Sample B: } \overline{X} = \frac{\Sigma X}{N} = \frac{30 + 40 + 50}{3} = \frac{120}{3} = 40$$

(b) Arrange the data into a table. Compute the deviations from $\overline{X}$ for each Sample and then square the deviations:

Sample A		
$\overline{X} = 50$		
X	$X - \overline{X}$	$(X - \overline{X})^2$
20	−30	900
80	+30	900
50	0	0
$\Sigma(X - \overline{X})^2 = 1800$		

Sample B		
$\overline{X} = 40$		
X	$X - \overline{X}$	$(X - \overline{X})^2$
30	−10	100
40	0	0
50	+10	100
$\Sigma(X - \overline{X})^2 = 200$		

(c) Calculate the degrees of freedom for each Sample:

Degrees of freedom = Sample size minus one

Sample A: $N - 1 = (3 - 1) = 2$

Sample B: $N - 1 = (3 - 1) = 2$

Total degrees of freedom = $\Sigma(N - 1) = 2 + 2 = 4$

(d) Compute the "within" Sample variance:

$$\sigma_X^2 = \frac{\text{Sum of all the squared deviations from } \overline{X}}{\text{Total number of degrees of freedom}}$$

$$\sigma_X^2 = \frac{\Sigma(X_1 - \overline{X})^2 + \Sigma(X_2 - \overline{X})^2}{(N_1 - 1) + (N_2 - 1)}$$

$$\sigma_X^2 = \frac{1800 + 200}{2 + 2} = \frac{2000}{4} = 500$$

Step 2. Calculate the "between" mean variance:

$$\sigma_{\overline{X}}^2 = \frac{\text{Sum of the squared deviations from } \overline{\overline{X}}}{\text{Number of degrees of freedom}}$$

$$\sigma_{\overline{X}}^2 = \frac{\Sigma(X - \overline{X})^2}{(N - 1)}$$

Needed:

$\overline{\overline{X}}$ (average of the means of the Samples), $\overline{X} - \overline{\overline{X}}$, $(\overline{X} - \overline{\overline{X}})^2$, $\Sigma(\overline{X} - \overline{\overline{X}})^2$, and the number of degrees of freedom.

(a) Calculate the average of the Sample means:

Sample A: $\overline{X} = 50$

Sample B: $\overline{X} = 40$

$$\overline{\overline{X}} = \frac{\Sigma\overline{X}}{N} = \frac{50 + 40}{2} = \frac{90}{2} = 45$$

(b) Arrange the data into a table. Compute the deviations from $\overline{X}$ for each Sample mean and then square the deviations:

Average of the Sample Means = $\overline{\overline{X}} = 45$		
Sample mean	Deviation	Deviation squared
$(\overline{X})$	$(\overline{X} - \overline{\overline{X}})$	$(\overline{X} - \overline{\overline{X}})^2$
50	+5	25
40	−5	25
$\Sigma(\overline{X} - \overline{\overline{X}})^2$	=	50

(c) Calculate the degrees of freedom:

Degrees of freedom = (number of Samples minus one) = $(2 - 1) = 1$

(d) Compute the "between" Sample variance:

$$\sigma_{\overline{X}}^2 = \frac{\text{Sum of the squared deviations from } \overline{\overline{X}}}{\text{Number of degrees of freedom}}$$

$$\sigma_{\overline{X}}^2 = \frac{50}{1} = 50$$

Step 3. Compare the "within" Sample variance to the between sample variance by computing the ratio between them:

$$F = \frac{\text{between sample variance}}{\text{within sample variance}} = \frac{50}{500} = .1$$

Step 4. Compare the computed ratio of .1 to the F table value at the between Sample degrees of freedom equals one, and the within Sample degrees of freedom = 4. Look up Column 1, row 4 in table:

From the table, the allowable F ratio is = 7.7. Since the computed ratio is less than 7.7, there is no significant difference between the two means.

PROBLEM 2: Determine if the means of the following two Samples are significantly different:

Sample A	Sample B
2	2
4	4
6	3

Solution

Step 1. Calculate the within Sample variance:

$$\sigma_X^2 = \frac{\Sigma(X - \overline{X})^2}{\Sigma(N-1)}$$

(a) Find the average means for each of the Samples:

Sample A: $\overline{X} = \frac{\Sigma X}{N} = \frac{2+4+6}{3} = \frac{12}{3} = 4$

Sample B: $\overline{X} = \frac{\Sigma X}{N} = \frac{2+4+3}{3} = \frac{9}{3} = 3$

(b) Arrange the data into a table. Compute the deviations from $\overline{X}$ for each Sample and then square the deviations:

Sample A		
$\overline{X} = 4$		
X	$X - \overline{X}$	$(X - \overline{X})^2$
2	-2	4
4	0	0
6	+2	4
$\Sigma(X - \overline{X})^2 = 8$		

Sample B		
$\overline{X} = 3$		
X	$X - \overline{X}$	$(X - \overline{X})^2$
2	-1	1
4	+1	1
3	0	0
$\Sigma(X - \overline{X})^2 = 2$		

(c) Calculate the degrees of freedom: for each Sample:

Degrees of freedom = Sample size minus one:

Sample A: N − 1 = (3 − 1) = 2

Sample B: N − 1 = (3 − 1) = 2

Total degrees of freedom: $= \Sigma(N-1)$

$= 2 + 2 = 4$

(d) Compute the "with in" Sample variance:

$$\sigma_X^2 = \frac{\text{Sum of all squared deviations from } \overline{\overline{X}}}{\text{Total number of degrees of freedom}}$$

$$\sigma_X^2 = \frac{\Sigma(X_1 - \overline{X})^2 + \Sigma(X_2 - \overline{X})^2}{(N_1 - 1) + (N_2 - 1)}$$

$$\sigma_X^2 = \frac{8+2}{2+2} = \frac{10}{4} = 2.5$$

Step 2. Calculate the "between" mean variance:

$$\sigma_{\overline{X}}^2 = \frac{\text{Sum of all squared deviations from } \overline{\overline{X}}}{\text{Number of degrees of freedom}}$$

$$\sigma_{\overline{X}}^2 = \frac{\Sigma(\overline{X} - \overline{\overline{X}})^2}{(N-1)}$$

(a) Calculate the average of the Sample means:

Sample A: $\overline{X} = 4$

Sample B: $\overline{X} = 3$

$$\overline{\overline{X}} = \frac{\Sigma\overline{X}}{N} = \frac{4+3}{2} = \frac{7}{2} = 3.2$$

(b) Arrange the data into a table. Compute the deviations from $\overline{X}$ for each Sample mean and then square the deviations:

Average of the Sample Means =		
$\overline{X} = 3.5$		
Sample mean	Deviation	Deviation squared
$(\overline{X})$	$(\overline{X} - \overline{\overline{X}})$	$(\overline{X} - \overline{\overline{X}})^2$
3	− .5	.25
4	+ .5	.25
$\Sigma(\overline{X} - \overline{\overline{X}})^2$	=	.50

(c) Calculate the degrees of freedom:

Degrees of freedom = (number of Samples minus one) = (2 – 1) = 1

(d) Compute the between Sample variance:

$$\sigma_{\overline{X}}^2 = \frac{\text{Sum of all squared deviations from } \overline{\overline{X}}}{\text{Number of degrees of freedom}}$$

$$\sigma_{\overline{X}}^2 = \frac{.50}{1} = .50$$

Step 3. Compare the within Sample variance to the between Sample variance by comparing the ratio between them:

$$F = \frac{\text{between Sample variance}}{\text{within Sample variance}} = \frac{.50}{2.5} = .2$$

Step 4. Compare the computed ratio of .2 to the F table value at the between Sample degrees of freedom equals one, and the within Sample degrees of freedom = 4. (Column 1, row 4 in the table):

From the table, the allowable F ratio is 7.7. Since the computed ratio is less than 7.7, there is no significant difference between the two means.

PROBLEM 3: Determine if the means of the following five Samples are significantly different:

Sample A	Sample B	Sample C	Sample D	Sample E
1	1	2	5	6
4	6	3	5	7
5	5	2	5	8
2	4	1	5	3

Solution

Step 1. Calculate the within sample variance:

$$\sigma_X^2 = \frac{\Sigma(X - \overline{X})^2}{\Sigma(N - 1)}$$

(a) Find the average means for each of the Samples:

Sample A: $\overline{X} = \frac{\Sigma X}{N} = \frac{1+4+5+2}{4} = \frac{12}{4} = 3$

Sample B: $\overline{X} = \frac{\Sigma X}{N} = \frac{1+6+5+4}{4} = \frac{16}{4} = 4$

Sample C: $\overline{X} = \frac{\Sigma X}{N} = \frac{2+3+2+1}{4} = \frac{8}{4} = 2$

Sample D: $\overline{X} = \frac{\Sigma X}{N} = \frac{5+5+5+5}{4} = \frac{20}{4} = 5$

Sample E: $\overline{X} = \frac{\Sigma X}{N} = \frac{6+7+8+3}{4} = \frac{24}{4} = 6$

(b) Arrange the data into a table, compute the deviations from X for each Sample and then square the deviations:

Sample A		
$\overline{X} = 3$		
X	$X - \overline{X}$	$(X - \overline{X})^2$
1	–2	4
4	1	1
5	2	4
2	–1	1
$\Sigma(X - \overline{X})^2 = 10$		

Sample B		
$\overline{X} = 4$		
X	$X - \overline{X}$	$(X - \overline{X})^2$
1	–3	9
6	2	4
5	1	1
4	0	0
$\Sigma(X - \overline{X})^2 = 14$		

Sample C		
$\overline{X} = 2$		
X	$X - \overline{X}$	$(X - \overline{X})^2$
2	0	0
3	1	1
2	0	0
1	–1	1
$\Sigma(X - \overline{X})^2 = 2$		

Sample D		
$\overline{X} = 5$		
X	$X - \overline{X}$	$(X - \overline{X})^2$
5	0	0
5	0	0
5	0	0
5	0	0
$\Sigma(X - \overline{X})^2 = 0$		

Sample E		
$\overline{X} = 6$		
X	$X - \overline{X}$	$(X - \overline{X})^2$
6	0	0
7	1	1
8	2	4
3	–3	9
$\Sigma(X - \overline{X})^2 = 14$		

(c) Calculate the degrees of freedom for each Sample:

Degrees of freedom = (Sample size minus 1)

Sample A; (N – 1) = (4 – 1) = 3

Sample B; (N – 1) = (4 – 1) = 3

Sample C: (N – 1) = (4 – 1) = 3

Sample D; (N – 1) = (4 – 1) = 3

Sample E; (N – 1) = (4 – 1) = 3

Total degrees of freedom = Σ(N – l) = (3 + 3 + 3 + 3 + 3) = 15.

(d) Compute the "within" Sample variance:

$$\sigma_X^2 = \frac{\text{Sum of all the squared deviations from } \overline{X}}{\text{Total number of degrees of freedom}}$$

$$\sigma_X^2 = \frac{\Sigma(X-\overline{X})_A^2 + \Sigma(X-\overline{X})_B^2 + \Sigma(X-\overline{X})_C^2}{(N_A-1)+(N_B-1)+(N_C-1)}$$

$$\frac{+\Sigma(X-\overline{X})_D^2 + \Sigma(X-\overline{X})_E^2}{+(N_D-1)+(N_E-1)}$$

$$\sigma_X^2 = \frac{10+14+2+0+14}{3+3+3+3+3} = \frac{40}{15} = 2.67$$

Step 2. Calculate the "between" mean variance:

$$\sigma_{\overline{X}}^2 = \frac{\text{Sum of all squared deviations from } \overline{\overline{X}}}{\text{Number of degrees of freedom}}$$

$$\sigma_{\overline{X}}^2 = \frac{\Sigma\left(\overline{X}-\overline{\overline{X}}\right)^2}{(N-1)}$$

(a) Calculate the average of the Sample means:

Sample A; $\overline{X} = 3$

Sample B; $\overline{X} = 4$

Sample C; $\overline{X} = 2$

Sample D; $\overline{X} = 5$

Sample E; $\overline{X} = 6$

$$X = \frac{\Sigma X}{N} = \frac{3+4+2+5+6}{5} = \frac{20}{5} = 4$$

(b) Arrange the data into a table, compute the deviations from $\overline{X}$ for each Sample mean and then square the deviations:

Average of the Sample Means $\overline{X} = 4$		
Sample mean	Deviation	Deviation squared
$(\overline{X})$	$(\overline{X}-\overline{\overline{X}})$	$(\overline{X}-\overline{\overline{X}})^2$
3	-1	1
4	0	0
2	-2	4
5	1	1
6	2	4
$\Sigma(\overline{X}-\overline{\overline{X}})^2$	=	10

(c) Calculate the degrees of freedom:

Degrees of freedom = (number of Samples minus 1) = (5 − 1) = 4

(d) Compute the "between" Sample variance:

$$\sigma_{\overline{X}}^2 = \frac{\text{Sum of the squared deviations from } \overline{\overline{X}}}{\text{Number of degrees of freedom}}$$

$$\sigma_{\overline{X}}^2 = \frac{10}{4} = 2.5$$

Step 3. Compare the "within" Sample variance to the between sample variance by calculating the ratio between them:

$$F = \frac{\text{between Sample variance}}{\text{within Sample variance}} = \frac{2.5}{2.66} = .94$$

Step 4. Compare the computed ratio of .94 to the F table value at the between Sample degrees of freedom = 4, and the within Sample degrees of freedom = 15 (Column 4, row .00 in the table):

From the table, the allowable F ratio is 2.4. Since the computed ratio is less than 2.4, there is no significant difference between the means.

15 COEFFICIENT OF CORRELATION

HOW TO DETERMINE THE EXTENT OF THE RELATIONSHIP BETWEEN TWO FACTORS

INTRODUCTION TO THE COEFFICIENT OF CORRELATION. The coefficient of correlation is used to determine whether there is some relationship or correlation between two factors, such as for example, the relationship between height and weight, or cigarette smoking and lung cancer, or age and blood pressure, or automobile mileage and accidents, etc.

By correlation, we mean that when we know the value of one factor, we can then in some way predict the value of the other factor. For example, when a doctor knows his patient's age, he can then predict what that person's normal blood pressure should be.

A coefficient of correlation is used to measure the size of the correlation between two factors. Thus, a coefficient equal to one means that there is a perfect or direct correlation or relationship, whereas, a coefficient equal to zero means that there is no correlation whatsoever between the factors.

An example of a coefficient of correlation equal to one is a mathematical formula. For example, the circumference of a circle is always equal to $2\pi R$. The letter R is the radius and π is a constant equal to 3.14. Therefore, the circumference of a circle is a direct function of its radius. Thus, the coefficient of correlation for the relationship between the circumference of a circle and the radius is equal to one.

An example of a zero coefficient of correlation is the relation between the annual rainfall in New York City to the population of Uganda. Here one factor has no effect whatsoever on the other.

A positive correlation coefficient means that there exists a direct correlation, that is, when one factor increases in size, so does the other factor change in some correlating way.

A negative correlation coefficient means that there exists an inverse correlation, that is, as one factor increases in size, the other factor decreases in size proportionately.

1. FORMULA FOR THE COEFFICIENT OF CORRELATION.
The formula for the coefficient of correlation is:

$$r = \frac{\Sigma(XY) - \frac{(\Sigma X)(\Sigma Y)}{N}}{\sqrt{\Sigma X^2 - \frac{(\Sigma X)^2}{N}} \cdot \sqrt{\Sigma Y^2 - \frac{(\Sigma Y)^2}{N}}}$$

r = Symbol for the coefficient of correlation.

N = The number of X's or the number of Y's.

X^2 = An individual number from one of the factors (factor X).

ΣX = Sum of all the individual numbers in factor X

$(\Sigma X)^2$ = The sum of all the individual numbers in factor X multiplied by itself

X^2 = An individual number in factor X multiplied by itself

ΣX^2 = Sum of all the squared individual numbers in factor X

Y = An individual number from one of the factors (factor Y)

ΣY = Sum of all the individual numbers in factor Y

$(\Sigma Y)^2$ = The sum of all the individual numbers in factor Y multiplied by itself

Y^2 = An individual number in factor Y multiplied by itself

ΣY^2 = Sum of all the squared individual numbers in factor Y

XY = A number in factor X multiplied by a number in factor Y

ΣXY = Sum of all the XY's

$(\Sigma X)(\Sigma Y)$ = Sum of all the X's multiplied by the sum of all the Y's

2. PROCEDURE FOR CALCULATING COEFFICIENT OF CORRELATION. The procedure for calculating the correlation coefficient can best be illustrated through the solution of a sample problem:

EXAMPLE: Find the coefficient of correlation for the following two sets of numbers:

X	Y
3	4
6	3
7	2
9	4

Solution

The coefficient of correlation =

$$r = \frac{\Sigma(XY) - \frac{(\Sigma X)(\Sigma Y)}{N}}{\sqrt{\Sigma X^2 - \frac{(\Sigma X)^2}{N}} \cdot \sqrt{\Sigma Y^2 - \frac{(\Sigma Y)^2}{N}}}$$

Needed: ΣX, $(\Sigma X)^2$, X^2, ΣX^2, ΣY, $(\Sigma Y)^2$, Y^2, ΣY^2, XY, ΣXY, and $(\Sigma X)(\Sigma Y)$

The easiest way of finding these values is by setting up a table as shown below:

X	Y	XY	X^2	Y^2
Col. 1	Col. 2	Col. 3	Col. 4	Col. 5
3	4	12	9	16
6	3	18	36	9
7	2	14	49	4
9	4	36	81	16
Totals				
25	13	80	175	45

Column 3 (XY) = Column 1 (X), multiplied by Column 2 (Y)

Column 4 (X^2) = Column 1 (X), squared

Column 5 (Y^2) = Column 2 (Y), squared

(Multiply across in the table.)

ΣX = Sum of Column 1 = 25

ΣY = Sum of Column 2 = 13

$\Sigma(XY)$ = Sum of Column 3 = 80

ΣX^2 = Sum of Column 4 = 175

ΣY^2 = Sum of Column 5 = 45

$(\Sigma X)^2$ = Sum of Column 1 squared = 25 × 25 = 625

$(\Sigma Y)^2$ = Sum of Column 2 squared = 13 × 13 = 169

N = Number of X's or the number of Y's = 4 The number of X's must equal the number of Y's.

$$r = \frac{80 - \frac{(25)(13)}{4}}{\sqrt{175 - \frac{(25)^2}{4}} \cdot \sqrt{45 - \frac{(13)^2}{4}}}$$

$$r = \frac{80 - 81.25}{\sqrt{175 - 156} \cdot \sqrt{45 - 42}}$$

$$r = \frac{-1.25}{\sqrt{19} \cdot \sqrt{3}} = \frac{-1.25}{(4.4)\ (1.7)}$$

$$r = \frac{-1.25}{7.5} = -.167$$

r is a very low number, being close to zero. Therefore, there is no correlation between the X and Y values in the data. The minus sign means an inverse correlation, i.e., as X increases Y decreases. This we can see from the data.

3. SAMPLE COEFFICIENT OF CORRELATION PROBLEMS.

PROBLEM 1: Find the coefficient of correlation for the following two sets of numbers:

X	Y
2	12
4	4
6	8

Solution

The coefficient of correlation = r

$$r = \frac{\Sigma(XY) - \frac{(\Sigma X)(\Sigma Y)}{N}}{\sqrt{\Sigma X^2 - \frac{(\Sigma X)^2}{N}} \cdot \sqrt{\Sigma Y^2 - \frac{(\Sigma Y)^2}{N}}}$$

Step 1. Set up a table for the required values:

X	Y	XY	X^2	Y^2
Col. 1	Col. 2	Col. 3	Col. 4	Col. 5
2	12	24	4	144
4	4	16	16	16
6	8	48	36	64
Totals				
12	24	88	56	224

ΣX = Sum of Column 1 = 12

ΣY = Sum of Column 2 = 24

$\Sigma(XY)$ = Sum of Column 3 = 88

ΣX^2 = Sum of Column 4 = 56

ΣY^2 = Sum of Column 5 = 224

$(\Sigma X)^2$ = Sum of Column 1 squared = 144

$(\Sigma Y)^2$ = Sum of Column 2 squared = 576

N = Number of X's or the number of Y's = 3

Step 2.

$$r = \frac{88 - \frac{(12)(24)}{3}}{\sqrt{56 - \frac{144}{3}} \cdot \sqrt{224 - \frac{576}{3}}}$$

$$r = \frac{88 - 96}{\sqrt{56 - 48} \cdot \sqrt{224 - 192}}$$

$$r = \frac{-8}{\sqrt{8} \cdot \sqrt{32}} = \frac{-8}{(2.8)(5.7)}$$

$$r = \frac{-8}{16} = -.5$$

The minus sign means an inverse correlation — as X goes up Y goes down in value.

PROBLEM 2: Find the coefficient of correlation for the following two sets of numbers:

X	Y
2	4
4	8
6	12

Solution

The coefficient of correlation equals:

$$r = \frac{\Sigma(XY) - \frac{(\Sigma X)(\Sigma Y)}{N}}{\sqrt{\Sigma X^2 - \frac{(\Sigma X)^2}{N}} \cdot \sqrt{\Sigma Y^2 - \frac{(\Sigma Y)^2}{N}}}$$

Step 1. Set up a table for the required values:

X	Y	XY	X^2	Y^2
Col. 1	Col. 2	Col. 3	Col. 4	Col. 5
2	4	8	4	16
4	8	32	16	64
6	12	72	36	144
Totals				
12	24	112	56	224

ΣX = Sum of Column 1 = 12

ΣY = Sum of Column 2 = 24

$\Sigma(XY)$= Sum of Column 3 = 112

ΣX^2 = Sum of Column 4 = 56

ΣY^2 = Sum of Column 5 = 224

$(\Sigma X)^2$ = Sum of Column 1 squared = 144

$(\Sigma Y)^2$ = Sum of Column 2 squared = 576

N = Number of X's or the number of Y's = 3

Step 2.

$$r = \frac{112 - \frac{(12)(24)}{3}}{\sqrt{56 - \frac{144}{3}} \cdot \sqrt{224 - \frac{576}{3}}}$$

$$r = \frac{112 - 96}{\sqrt{56 - 48} \cdot \sqrt{224 - 192}}$$

$$r = \frac{16}{\sqrt{8} \cdot \sqrt{32}} = \frac{16}{(2.8)(5.7)}$$

$$r = \frac{16}{16} = 1$$

There is a perfect correlation between X and Y. From the data in the problem you can see that Y = 2X.

PROBLEM 3: Is there correlation between the following two sets of numbers?

X	Y
1	3
3	2
2	6
1	4

Solution

The coefficient of correlation is equal to:

$$r = \frac{\Sigma(XY) - \frac{(\Sigma X)(\Sigma Y)}{N}}{\sqrt{\Sigma X^2 - \frac{(\Sigma X)^2}{N}} \cdot \sqrt{\Sigma Y^2 - \frac{(\Sigma Y)^2}{N}}}$$

Step 1. Set up a table for the required values:

X	Y	XY	X^2	Y^2
Col. 1	Col. 2	Col. 3	Col. 4	Col. 5
1	3	3	1	9
3	2	6	9	4
2	6	12	4	36
1	4	4	1	16
Totals				
7	15	25	15	65

ΣX = Sum of Column 1 = 7

ΣY = Sum of Column 2 = 15

$\Sigma(XY)$ = Sum of Column 3 = 25

ΣX^2 = Sum of Column 4 = 15

ΣY^2 = Sum of Column 5 = 65

$(\Sigma X)^2$ = Sum of Column 1 squared = 49

$(\Sigma Y)^2$ = Sum of Column 2 squared = 225

N = Number of X's or the number of Y's = 4

Step 2.

$$r = \frac{25 - \frac{(7)(15)}{4}}{\sqrt{15 - \frac{49}{4}} \cdot \sqrt{65 - \frac{225}{4}}}$$

$$r = \frac{-1.3}{\sqrt{2.8} \cdot \sqrt{8.8}} = \frac{-1.3}{(1.67)(2.97)}$$

$$r = \frac{-1.2}{5.0} = -.24$$

The computed value of r is close to zero, therefore there is no correlation between X and Y.

PROBLEM 4: Using the following data taken over a 5-month period, determine if there is a correlation between monthly rainfall (X) and the height in inches (Y) for a particular garden plant:

X	Y
1	3
3	5
5	7
6	9
7	9

Solution

The coefficient of correlation is equal to:

$$r = \frac{\Sigma(XY) - \frac{(\Sigma X)(\Sigma Y)}{N}}{\sqrt{\Sigma X^2 - \frac{(\Sigma X)^2}{N}} \cdot \sqrt{\Sigma Y^2 - \frac{(\Sigma Y)^2}{N}}}$$

Step 1. Set up a table for the required values:

X	Y	XY	X^2	Y^2
Col. 1	Col. 2	Col. 3	Col. 4	Col. 5
1	3	3	1	9
3	5	15	9	25
5	7	35	25	49
6	9	54	36	81
7	9	63	49	81
Totals				
22	33	170	120	245

ΣX = Sum of Column 1 = 22

ΣY = Sum of Column 2 = 33

$\Sigma(XY)$ = Sum of Column 3 = 170

ΣX^2 = Sum of Column 4 = 120

ΣY^2 = Sum of Column 5 = 245

$(\Sigma X)^2$ = Sum of Column 1 squared = 484

$(\Sigma Y)^2$ = Sum of Column 2 squared = 1089

N = Number of X's or number of Y's = 5

Step 2.

$$r = \frac{170 - \frac{(22)(33)}{5}}{\sqrt{120 - \frac{484}{5}} \cdot \sqrt{245 - \frac{1089}{5}}}$$

$$r = \frac{170 - 145.2}{\sqrt{120 - 97} \cdot \sqrt{245 - 218}}$$

$$r = \frac{24.8}{\sqrt{23} \cdot \sqrt{27}} = \frac{24.8}{(4.8)(5.2)}$$

$$r = \frac{24.96}{25} = .998 \text{ or } 1$$

The value is close to 1, therefore there is an excellent correlation between X and Y. From the data you can see that Y = X + 2

There is only one point which is off and that is X = 6, Y = 9

16 CHI SQUARE TEST

HOW TO DETERMINE IF AN ACTUAL FREQUENCY DIFFERS SIGNIFICANTLY FROM THE EXPECTED FREQUENCY

INTRODUCTION TO THE CHI SQUARE TEST. Chi Square is pronounced keye square and it is represented by the symbol χ^2.

The Chi Square Test is a technique for determining if the actual frequency varies significantly from a theoretical or expected frequency.

Frequency, you will remember, was simply defined as the number of times that a given item occurs in the data.

The theoretical frequency is the frequency which would be expected if the data conformed to a theoretical distribution, such as the normal, binomial, Poisson, etc.

1. FORMULA FOR CHI SQUARE. The formula for calculating chi square is:

$$\chi^2 = \Sigma\frac{(Fo - Fe)^2}{Fe}$$

Fe = Theoretical or expected frequency

Fo = Actual or observed frequency

(Fo – Fe) = The individual difference between the actual and corresponding theoretical frequency in the data

$(Fo – Fe)^2$ = Square of the individual difference in frequency

When the actual frequency is equal to its corresponding theoretical frequency, then $\chi^2 = 0$

The greater the difference between the actual and the theoretic-cal frequencies, the larger the value of χ^2 will be.

2. INTRODUCTION TO THE CHI SQUARE TABLE. In order to determine if the computed value of χ^2 is considered significant, it is compared to a table of χ^2 values. The chi square table contains the critical values of χ^2 at different confidence or probability of occurrence levels.

   For example, the 5% confidence level means that the χ^2 value in the table will be exceeded in 5 out of 100 times. The 1% level will be exceeded in 1 out of 100 times.

The difference between the actual and theoretical frequency is considered significant when the computed value of χ^2 is greater than the table value of χ^2 at the P = .05 and P = .01 significance levels.

The chi square table values are also tabulated for a series of degrees of freedom. The degrees of freedom are equal to the number of categories in the data, minus one.

3. CHI SQUARE TABLE. Shown below is a two-decimal place chi square table for certain selected confidence or probability of occurrence levels.

Table of Values of χ^2

Degrees of Freedom	Selected Confidence Levels			
	P = .99	P = .95	P = .05	P = .01
1	.0001	.004	3.85	6.64
2	.02	.10	5.99	9.21
3	.12	.35	7.82	11.35
4	.30	.70	9.49	13.28
5	.55	1.15	11.07	15.09
6	.87	1.64	12.59	16.81
7	1.24	2.17	14.07	18.48
8	1.65	2.73	15.51	20.09
9	2.09	3.33	16.92	21.67
10	2.56	3.94	18.31	23.21

4. PROCEDURE FOR CALCULATING χ^2. The procedure for calculating chi square can best be illustrated through the solution of a sample problem.

EXAMPLE: In an experiment, a coin was tossed 1,000 times. Heads appeared 550 times and tails 450 times. We expected to get 500 heads and 500 tails, since the probability of getting heads is .5 and the probability of tails is .5. Does the result indicate a biased coin?

Solution

Step 1. Arrange the data of the problem into a simple table for ease of analysis:

Result of Toss	Actual Frequency	Theoretical Frequency
	Fo	Fe
Heads	550	500
Tails	450	500
Total	1000	1000

Step 2.

$$\chi^2 = \Sigma \frac{(Fo - Fe)^2}{Fe}$$

$$\chi^2 = \chi^2 \text{ Heads} + \chi^2 \text{ Tails}$$

$$\chi^2 = \underset{\text{Heads}}{\frac{(Fo - Fe)^2}{Fe}} + \underset{\text{Tails}}{\frac{(Fo - Fe)^2}{Fe}}$$

$$\chi^2 = \frac{(550 - 500)^2}{500} + \frac{(450 - 500)^2}{500}$$

$$\chi^2 = \frac{(50)^2}{500} + \frac{(-50)^2}{500}$$

$$\chi^2 = \frac{2500}{500} + \frac{2500}{500}$$

$$\chi^2 = 5 + 5 = 10$$

Step 3. There are two categories: heads and tails. The degrees of freedom are therefore equal to: (2 – 1) = 1

Step 4. To determine if $\chi^2 = 10$ is significant, we have to compare it to the chi square table values:

From the table, at one degree of freedom, chi square will be greater than 3.85 in 1 out of 200 times (the .5% level). (In the table look in row 1, P = .05 column),

Since chi square was calculated as being equal to 10, we can assume that the coin was biased. Once the result obtained will probably occur in less than 1 out of 200 times.

5. SAMPLE CHI SQUARE PROBLEMS.

PROBLEM 1: In evaluating grinding wheels for a machine shop, records were kept for six weeks on wheels made by five different manufacturers A, B, C, D and E.

Each wheel was expected to last equally long. The breakage results obtained were the following: A = 9, B = 10, C = 12, D = 11, and E = 8 breakages.

Solution

Step 1. Arrange the data into a table:

Manufacturer	Observed Number of Breakages	Expected Number of Breakages
	Fo	Fe
A	9	10
B	10	10
C	12	10
D	11	10
E	8	10
Total	50	50

Since each wheel was expected to last equally as long, the expected breakage (Fe) was obtained by adding up the actual number of breakages and then dividing them equally among all the manufacturers.

Step 2.

$$\chi^2 = \Sigma \frac{(Fo - Fe)^2}{Fe}$$

$$\chi^2 = \chi_A{}^2 + \chi_B{}^2 + \chi_C{}^2 + \chi_D{}^2 + \chi_E{}^2$$

$$\chi^2 = \frac{(9-10)^2}{10} + \frac{(10-10)^2}{10} + \frac{(12-10)^2}{10} + \frac{(11-10)^2}{10} + \frac{(8-10)^2}{10}$$

$$\chi^2 = \frac{(-1)^2}{10} + \frac{(0)^2}{10} + \frac{(2)^2}{10} + \frac{(1)^2}{10} + \frac{(-2)^2}{10}$$

$$\chi^2 = \frac{1}{10} + \frac{0}{10} + \frac{4}{10} + \frac{1}{10} + \frac{4}{10}$$

$$\chi^2 = \frac{10}{10} = 1$$

Step 3. There are five categories, one for each manufacturer. The degrees of freedom are therefore equal to (5 – 1) = 4

Step 4. Compare $\chi^2 = 1$ to the table values for 4 degrees of freedom:

From the table at 4 degrees of freedom, chi square will be greater than .70 in 95 out of 100 times (the 95% level),

we, therefore, can assume that there is no significant difference between the grinding wheels.

PROBLEM 2: The following result was obtained when a single die was thrown 300 times. Number one had shown up 50 times, number two appeared 42 times, number three was 45 times, number four came up 60 times, number five was seen 51 times, and number six showed 52 times. If the die was unbiased, each number would have an equal chance of appearing. Therefore, the expected frequency of each number is 300 ÷ 6 = 50. Do the observed results differ significantly from the expected results?

Solution

Step 1. Arrange the data into a table:

Number Shown on Die	Observed Frequency Fo	Expected Frequency Fe
1	50	50
2	42	50
3	45	50
4	60	50
5	51	50
6	52	50
Totals	300	300

Step 2.

$$\chi^2 = \Sigma\frac{(Fo - Fe)^2}{Fe}$$

$$\chi^2 = \chi_1^2 + \chi_2^2 + \chi_3^2 + \chi_4^2 + \chi_5^2 + \chi_6^2$$

$$\chi^2 = \frac{(50-50)^2}{50} + \frac{(42-50)^2}{50} + \frac{(45-50)^2}{50}$$

$$+ \frac{(60-50)^2}{50} + \frac{(51-50)^2}{50} + \frac{(52-50)^2}{50}$$

$$\chi^2 = \frac{(0)^2}{50} + \frac{(-8)^2}{50} + \frac{(-5)^2}{50} + \frac{(10)^2}{50} + \frac{(1)^2}{50} + \frac{(2)^2}{50}$$

$$\chi^2 = \frac{0}{50} + \frac{64}{50} + \frac{25}{50} + \frac{100}{50} + \frac{1}{50} + \frac{4}{50}$$

$$\chi^2 = \frac{194}{50} = 3.88$$

Step 3. There are six categories, one for each number. The degrees of freedom are therefore equal to (6 – 1) = 5

Step 4. Compare χ^2 = 3.88 to the table values for 5 degrees of freedom.

From the table at 5 degrees of freedom, chi square will be greater than 1.15 in 95 out of 100 times (the 95% level).

We can therefore assume that the die is fair.

PROBLEM 3: A bowl contains 10 green balls, 20 blue balls, 30 white balls and 40 red balls. In an experiment, a ball is picked out of this bowl, its color is noted and then the ball is replaced in the bowl before picking up the next ball. This experiment was repeated until 200 balls were looked at, and the following distribution was obtained.

Color	Frequency
Green	20
Blue	30
White	60
Red	90

Do the observed results differ significantly from the expected results?

Solution

Step 1. The bowl contains 100 colored balls. 10% are green, 20% are blue, 30% are white, and 40% are red. Therefore the expected frequencies would be based on these same percentages for the 200 balls.

(a) Expected frequency for green balls = 200 × 10% = 20

(b) Expected frequency for the blue balls = 200 × 20% = 40

(c) Expected frequency for the white balls = 200 × 30% = 60

(d) Expected frequency for the red balls = 200 × 40% = 80

Total = 200

Arrange the data into a table:

Color	Observed Frequency Fo	Expected Frequency Fe
Green	20	20
Blue	30	40
White	60	60
Red	90	80
Totals	200	200

Step 2.

$$\chi^2 = \Sigma \frac{(Fo - Fe)^2}{Fe}$$

$$\chi^2 = \chi_1^2 + \chi_2^2 + \chi_3^2 + \chi_4^2$$

$$\chi^2 = \frac{(20-20)^2}{20} + \frac{(30-40)^2}{40} + \frac{(60-60)^2}{60} + \frac{(90-80)^2}{80}$$

$$\chi^2 = \frac{0}{20} + \frac{(-10)^2}{40} + \frac{0}{60} + \frac{(10)^2}{80}$$

$$\chi^2 = \frac{0}{20} + \frac{100}{40} + \frac{0}{60} + \frac{100}{80}$$

$$\chi^2 = \frac{200}{80} + \frac{100}{80} = \frac{300}{80} = 3.75$$

Step 3. There are four colors. The degrees of freedom are therefore equal to (4 – 1) = 3.

Step 4. Compare χ^2 = 3.75 to the table value for 3 degrees of freedom:

From the table at 3 degrees of freedom, chi square will be greater than .35 in 95 out of 100 times (the 95% level),

Therefore, there is no significant difference between the observed and expected frequencies.

17 LINEAR REGRESSION

HOW TO USE THE METHOD OF LEAST SQUARES TO FIT A STRAIGHT TREND LINE TO PLOTTED DATA

INTRODUCTION TO THE EQUATION FOR A STRAIGHT LINE. The general equation for a straight line is:

$Y = aX + b$

Y = Y axis location of a point on the straight line

X = X axis location of a point on the straight line

b = The Y axis location where the straight line intersects or would intersect the Y axis

a = Slope of the straight line

$$a = \frac{Y_2 - Y_1}{X_2 - X_1}$$

Y_2, X_2 and Y_1, X_1 are locations of two points on the straight line

a and b are therefore the parameters which define and locate a straight line

The following example illustrates a graphical interpretation of the parameters a and b.

EXAMPLE: Find the value of a and b for the straight line plotted on the graph below:

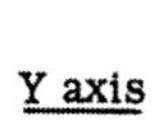

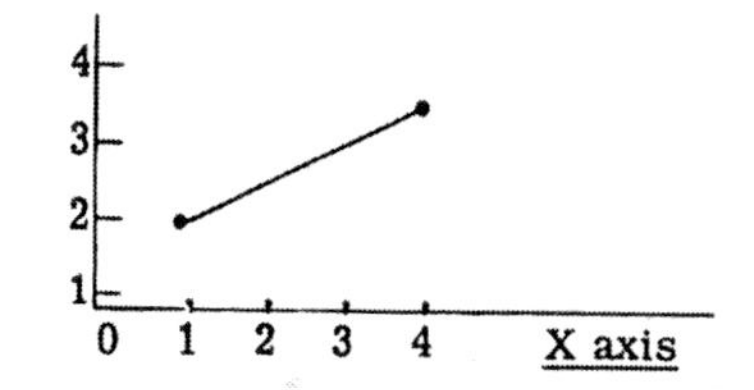

Solution

(1) b = Intersection of the straight line on the Y axis. Therefore, we extend the line until it intersects the Y axis

b = point of intersection = 1.6

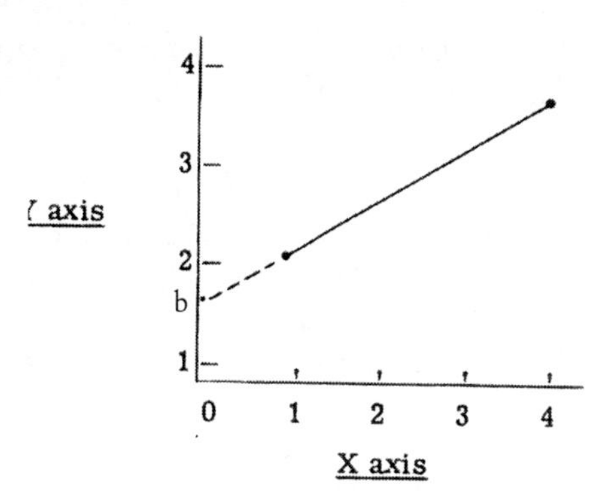

(2) a = Slope of the line = $\dfrac{Y_2 - Y_1}{X_2 - X_1}$

a = Change in height per given change in horizontal distance

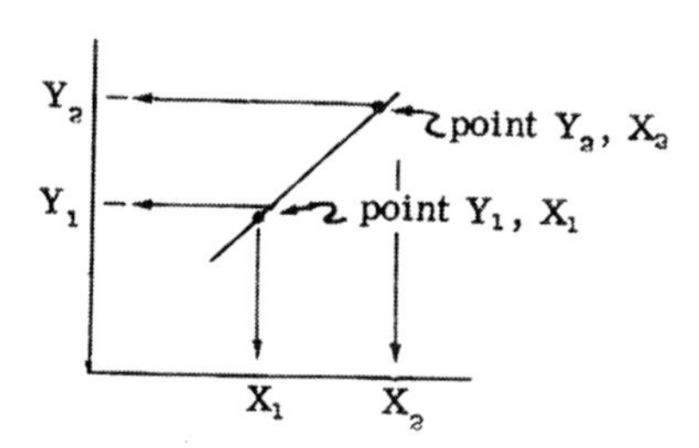

Y_2, X_2 and Y_1, X_1 can be the location of any two points on the straight line

$$a = \frac{Y_2 - Y_1}{X_2 - X_1} = \frac{4-2}{4-1} = \frac{2}{3}$$

There are three horizontal units (X) for every one unit of height (Y)

(3) With a = 1.5 and b = $\frac{2}{3}$ the equation for the straight line in this example is equal to:

$$Y = .67X + \frac{2}{3}X$$

That is, for any value of X on the straight line, we can find its corresponding Y value by solving the equation for Y. Conversely, if we know Y we can find X.

INTRODUCTION TO LINEAR REGRESSION. In actual practice data plotted on a graph is usually scattered. It rarely falls in a straight line. We may, however, want to represent the plotted data with a straight line for ease of analysis and to determine the trend of the data. For example, we may be interested in knowing if there is an upward trend, or a downward trend, and how steep is the trend for the relationship between two items such as: sales versus advertising expenditures, production quantity versus unit cost, automobile gas mileage versus speed, etc.

The term linear regression is used to describe a method of determining the equation of the straight line which best represents the trend of the data.

PROCEDURE FOR FITTING A STRAIGHT OR LINEAR REGRESSION LINE TO PLOTTED DATA BY THE METHOD OF LEAST SQUARES. In order to represent data with a straight trend line, we need to know the values of the parameters a and b in the general equation for a straight line $Y = aX + b$. These parameters can be found by setting up two simultaneous equations involving a and b, and then solving the equations. The procedure for doing this is the following:

Step 1. Since each plotted point has an X axis location and a Y axis location, we substitute the appropriate values of X and Y in the equation $Y = aX + b$, for each plotted point. There will be as many equations as there are paired X and Y values.

Step 2. Add up all the individual equations in Step 1 to produce a single "total" equation which we will call equation #1 = $\Sigma(Y = aX + b)$.

Step 3. Multiply all of the individual equations in Step 1 by their respective X value. The general expression for these series of equations is: $XY = aX^2 + bX$.

Step 4. Add up all the individual equations in Step 3 to produce a single total equation, which we will call equation #2 = $\Sigma\ (XY = aX^2 + bX)$.

Step 5. Solve the simultaneous equations #1 (Step 2) and #2 (Step 4) for the parameters a and b.

Step 6. Substitute the values for a and b from Step 5 in the equation for the straight line $Y = aX + b$. This resulting equation best represents the trend of the data.

The term least squares means that the trend line is drawn so that the sum of the squares of all the individual distances between the trend line and each plotted point is at a minimum value.

1. SAMPLE LINEAR REGRESSION PROBLEMS.

PROBLEM 1: Find the least squares trend line for the following data on annual sales versus year in production for a certain product.

Year in production =	#1	#2	#3	#4	#5
Annual sales in millions =	3	5	7	9	11

Solution

Step 1. Let sales = Y and year = X. X = fixed data, Y = variable data. We are going to substitute the values for X and Y in the equations:

$$Y = aX + b \text{ and}$$

$$XY = aX^2 + bX$$

This can best be done by using a table as shown below.

X	Y	$Y = aX + b$	$XY = aX^2 + bX$
Col. 1	Col. 2	Column 3	Column 4
1	3	$3 = a(l) + b$	$(1)(3) = a(1)(1) + b(1)$
2	5	$5 = a(2) + b$	$(2)(5) = a(2)(2) + b(2)$
3	7	$7 = a(3) + b$	$(3)(7) = a(3)(3) + b(3)$
4	9	$9 = a(4) + b$	$(4)(9) = a(4)(4) + b(4)$
5	11	$11 = a(5) + b$	$(5)(11) = a(5)(5) + 6(5)$

Step 2. Multiply the equations out, add Column 3 to get Equation #1 = $\Sigma\ (Y = aX + b)$ and add Column 4 to get Equation #2 = $\Sigma\ (XY = aX^2 + bX)$:

X	Y	$Y= aX + b$	$XY=aX^2 + bX$
Col.1	Col. 2	Column 3	Column 4
1	3	$3 = a + b$	$3 = 1a + 1b$
2	5	$5 = 2a + b$	$10 = 4a + 2b$
3	7	$7 = 3a + b$	$21 = 9a + 3b$
4	9	$9 = 4a + b$	$36 = 16a + 4b$
5	11	$11 = 5a + b$	$55 = 25a + 5b$
Totals:		$35 = 15a + 5b$	$125 = 55a + 15b$

Equation #1: $35 = 15a + 5b$

Equation #2: $125 = 55a + 15b$

Step 3. Solve the simultaneous Equations #1 and #2 for the parameters a and b:

(a) We have to subtract Equations #1 and #2 from each other in order to get rid of either the a or b term. This will leave us with one equation and one unknown:

If we multiply both sides of Equation #1 by 3, we would get a 15b term in this equation. Through subtraction, this 15b term in Equation #1 will cancel out the 15b term in Equation #2, leaving a term with an "a" in the resultant sum.

$$3(35) = 3(15a) + 3(5b)$$

$$105 = 45a + 15b$$

(b) Subtract Equation #1 from Equation #2 to eliminate the "b" term to solve for "a":

$$\begin{array}{rl} & 125 = 55a + 15b \\ - & \underline{105 = 45a + 15b} \\ & 20 = 10a + 0 \\ & a = \frac{20}{10} = 2 \end{array}$$

(c) Substitute a = 2 in Equation #1 and then solve this equation for b:

$$35 = 15(2) + 5b$$

$$35 = 30 + 5b$$

$$5b = 35 - 30 = 5$$

$$b = \frac{5}{5} = 1$$

Step 4. Substitute the computed values for a and b in the general equation for a straight line Y = aX + b. This, then, is the trend line for the original data of sales versus year in production:

$$Y = 2X + 1$$

Note: In this example, the trend line coincides with the plotted data, proving that this technique works.

Actual X	Actual Y	Calculated Y from Y = 2X + 1
1	3	Y = 2(1) + 1 = 3
2	5	Y = 2(2) + 1 = 5
3	7	Y = 2(3) + 1 = 7
4	9	Y = 2(4) + 1 = 9
5	11	Y = 2(5) + 1 =11

Step 5. Plot the original data on a graph. Then draw in the trend line using the actual X and the calculated Y:

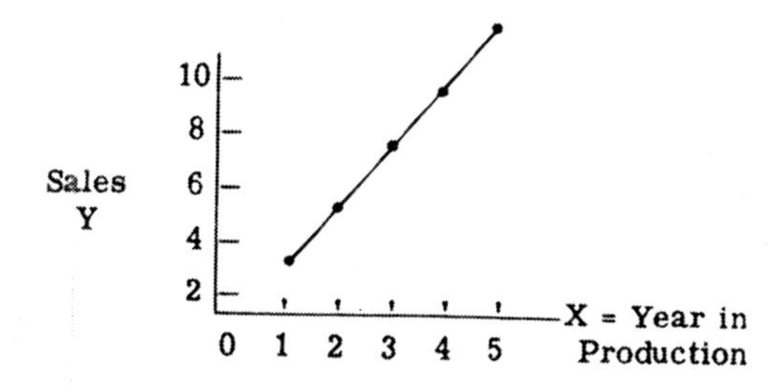

PROBLEM 2: Find the least squares trend line for the folio wing data on production quantity versus unit cost for a product:

Production (in thousands)	1	2	4	6
Unit Cost (in dollars)	5	4	3	2

Solution

Step 1. Let production = X and unit cost = Y. X = fixed data. Y = variable data. We are going to substitute the values for X and Y in the equations:

$$Y = aX + b$$

$$XY = aX^2 + bX$$

This can be done best by using a table as shown below:

X	Y	Y = aX + b	$XY = aX^2 + bX$
Col. 1	Col. 2	Col. 3	Col. 4
1	5	5 = a(1) + b	(1)(5) = a(1)(1) + b(1)
2	4	4 = a(2) + b	(2)(4) = a(2)(2) + b(2)
4	3	3 = a(4) + b	(4)(3) = a(4)(4) + b(4)
6	2	2 = a(6) + b	(6)(2) = a(6)(6) + b(6)

Step 2. Multiply the equations out, add Column #3 to get Equation #1 = Σ (Y aX + b), and then add Col #4 to get Equation #2 = Σ (XY aX^2 + bX):

X	Y	Y = aX + b	$XY = aX^2 + bX$
Col. 1	Col. 2	Column 3	Column 4
1	5	5 = a + b	5 = a + b
2	4	4 = 2a + b	8 = 4a + 2b
4	3	3 = 4a + b	12 = 16a + 4b
6	2	2 = 6a + b	12 = 36a + 6b
Totals		14 = 13a + 4b	37 = 57a + 13b

Equation #1: 14 = 13a + 4b

Equation #2: 37 = 57a + 13b

Step 3. Solve the simultaneous Equations #1 and #2 for the parameters a and b:

(a) We have to subtract Equations #1 and #2 from each other in order to get rid of either the a or b term,

if we multiply both sides of Equation #1 by 13 and both sides of Equation #2 by 4, we would end up with a 52b term in both equations. This term will then drop out through subtraction, leaving only a term with an "a" in the resultant sum:

#1: 13(14) = 13(13a) + 13(4b)

$$182 = 169a + 52b$$

#2: $4(37) = 4(57a) + 4(13b)$

$148 = 228a + 52b$

(b) Subtract Equation #2 from Equation #1:

$$182 = 169a + 52b$$
$$-\ \underline{148 = 228a + 52b}$$
$$34 = -59a + 0$$
$$a = \frac{34}{-59} = -.6 \text{ (approx.)}$$

(c) Substitute a = –.6 in Equation #1 and then solve this equation for b:

$$14 = 13(-.6) + 4b$$
$$14 = -7.8 + 4b$$
$$4b = 14 + 7.8 = 21.8$$
$$b = \frac{21.8}{4} = 5.5 \text{ (approx.)}$$

Step 4. Substitute the computed values for a + b in the general equation for a straight line Y = aX + b. This, then, is the trend line for the original data of production versus unit cost:

$$Y -.6_X + 5.5$$

Actual X	Actual Y	Calculated Y from Y = –.6X + 5.5
Production	Unit Cost	Calculated Unit Cost
1	5	Y = –.6(l) + 5.5 = 4.9
2	4	Y = –.6(2) + 5.5 = 4.3
4	3	Y = –.6(4) + 5.5 = 3.1
6	2	Y = –.6(6) + 5.5 = 1.9

Step 5. Plot the data on a graph. Draw in the trend line using the actual X and the calculated Y:

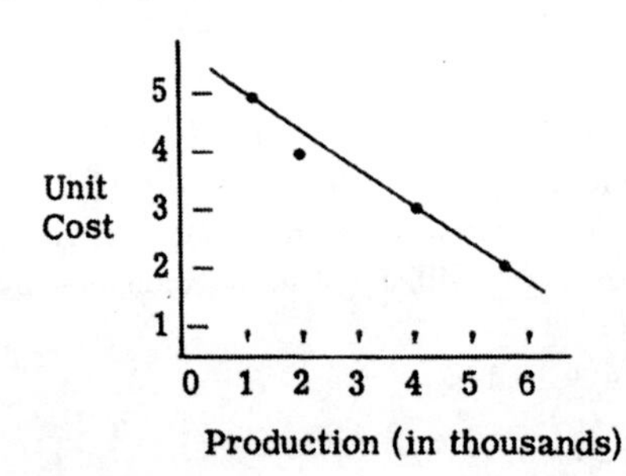

PROBLEM 3: Find the least squares trend for the following data on population versus year for a small town:

Year	1954	1955	1956	1957	1958
Years from 1954	0	1	2	3	4
Population (in thousands)	10	10	15	17	19
Year	1959	1960	1961	1962	1963
Years from 1954	5	6	7	8	9
Population (in thousands)	22	24	26	27	30

Solution

Step 1. Let year from 1954 = X, and population = Y, because X is fixed and Y is the variable:

we are going to substitute the values for X and Y in the equations:

$$Y = aX + b$$
$$XY = aX^2 + bX$$

This is shown in the following table:

X	Y	Y= aX + b	XY = aX² + bX
Col. 1	Col. 2	Column 3	Column 4
0	10	10 = a(0) + b	(0)(10) = a(0)(0) + b(0)
1	10	10 = a(l) + b	(1)(10) = a(l)(l) + b(l)
2	15	15 = a(2) + b	(2)(15) = a(2)(2) + b(2)
3	17	17 = a(3) + b	(3)(17) = a(3)(3) + b(3)
4	19	19 = a(4) + b	(4)(19) = a(4)(4) + b(4)
5	22	22 = a(5) +b	(5)(22) = a(5)(5) + b(5)
6	24	24 = a(6) + b	(6)(24) = a(6)(6) + b(6)
7	26	26 = a(7) + b	(7)(26) = a(7)(7) + b(7)
8	27	27 = a(8) + b	(8)(27) = a(8)(8) + b(8)
9	30	30 = a(9) + b	(9)(30) = a(9)(9) + b(9)

Step 2. Multiply the equations out, add Column 3 to get Equation #1 = $\Sigma(Y = aX + b)$, and add Column 4 to get Equation #2 = $\Sigma\,(XY= aX^2 + bX)$:

X	Y	Y = aX + b	XY = aX² + bX
Col. 1	Col. 2	Column 3	Column 4
0	10	10 = 0 + b	0 = 0 + b
1	10	10 = a + b	10 = a + b
2	15	15 = 2a + b	30 = 4a + 4b
3	17	17 = 3a + b	51 = 9a + 3b
4	19	19 = 4a + b	76 = 16a + 4b
5	22	22 = 5a + b	110 = 25a + 5b
6	24	24 = 6a + b	144 = 36a + 6b
7	26	26 = 7a + b	182 = 49a + 7b
8	27	27 = 8a + b	216 = 64a + 8b
9	30	30 = 9a + b	270 = 81a + 9b
Totals		200 = 45a + 10b	1089 = 285a + 45b

Equation #1: $\Sigma (Y = aX + b)$
Sum of Column 3 = 200 = 45a + 10b

Equation #2: $\Sigma (XY = aX^2 + bX)$
Sum of Column 4 = 1089 = 285a + 45b

Step 3. Solve simultaneous Equations #1 and #2 for the parameters a and b:

(A) We have to subtract Equations #1 and #2 from each other to get rid of either a or b term. This will leave one equation and one unknown:

If we multiply both sides of Equation #1 by 9, and both sides of Equation #2 by 2, we would get a 90b term in both equations. The 90b term will subtract out leaving only a term with an "a" in the result:

#1: 9(200) = 9(45a) + 9(10b)

1800 = 405a + 90b

#2: 2(1089) = 2(285a) + 2(45b)

2178 = 570a + 90b

(B) Subtract Equation #1 from Equation #2 to eliminate the "b" term and solve for "a":

2178 = 570a + 90b

− 1800 = 405a + 90b

378 = 165a + 0

$$a = \frac{378}{165} = 2.3$$

(C) Substitute a = 2.3 in Equation #1 and then solve this Equation for b:

200 = 45(2.3) + 10b

200 = 103 + 10b

10b = 200 − 103 = 97

$$b = \frac{97}{10} = 9.7$$

Step 4. Substitute the computed values for a and b in the general equation for a straight line Y = aX + b. This, then, is the trend line for the original data of population versus year:

Y = 2.3X + 9.7

Year	Actual Population	Calculated Y from Y = 2.3X + 9.7
0	10	Y = 2. 3 (0) + 9.7 = 9.7
1	10	Y = 2.3 (1) + 9.7 = 12.0
2	15	Y = 2.3 (2) + 9.7 = 14.3
3	17	Y = 2.3 (3) + 9.7 = 16.6
4	19	Y = 2.3 (4) + 9.7 = 18.9
5	22	Y = 2.3 (5) + 9.7 = 21.2
6	24	Y = 2. 3 (6) + 9.7 = 23.5
7	26	Y = 2.3 (7) + 9.7 = 25.8
8	27	Y = 2.3 (8) + 9.7 = 28.
9	30	Y = 2.3 (9) + 9.7 = 30.4

Step 5. Plot the data on the graph. Draw in the trend line using the actual X and the calculated Y:

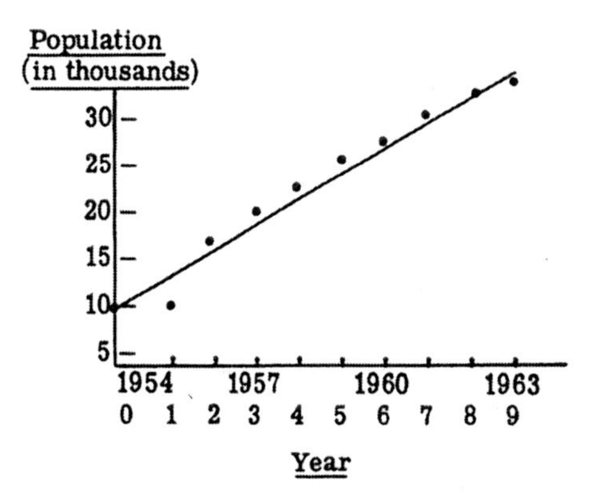

18 INDEX NUMBERS

HOW TO COMPUTE VARIOUS PRICE INDEXES

INTRODUCTION TO INDEX NUMBERS. An index number is generally used to measure the change in the price of some factor during two periods of time. It is simply the ratio of the price at one base period (the given period) divided by the price at the period it is being compared to. For example, if in 1950 a certain factory produced $15 million in product, in 1960 it produced $30 million, and in 1965 it produced $45 million, then using 1950 as the basis for comparison (base period), the indexes of this factory's production are:

YEAR	PRODUCTION	INDEX 1950 = 100
1950	$15 million	$\frac{15}{15} = 1.00$
1960	$30 million	$\frac{30}{15} = 2.00$
1965	$45 million	$\frac{45}{15} = 3.00$

We can see from the index column that the production in 1965 is 3 times the production in 1950 and 2 times the production in 1960.

SIMPLE AGGREGATIVE PRICE INDEX. A simple price index is equal to the sum of the given period prices divided by the sum of the base period prices. Expressed symbolically:

$$I = \frac{\Sigma p_n}{\Sigma p_o} \times 100$$

I = Index number

p_n = Given price period

Σp_n = Sum of the given period prices

p_o = Base period price (basis for comparison)

Σp_o = Sum of the base period prices

100 = Factor that converts the index number into a per cent.

1. SAMPLE SIMPLE AGGREGATIVE PRICE INDEX PROBLEMS.

PROBLEM 1: For the following items, use an index to compare their average 1965 prices to their average 1960 prices.

Item		1960 Average Price	1965 Average Price
Meat	(lb.)	$.65	$.75
Fish	(lb.)	.55	.60
Potatoes	(lb.)	.10	.15
Butter	(lb.)	.70	.80

Solution

$$I = \frac{\Sigma p_n}{\Sigma p_o} \times 100$$

$$I = \frac{.75 + .60 + .15 + .80}{.65 + .55 + .10 + .70} \times 100$$

$$I = \frac{2.30}{2.00} \times 100 = 1.15 \times 100$$

I = 115 or 115%

This means that the prices of these items increased 1.15 times from 1960 to 1965.

PROBLEM 2: The following table gives the average cost of three household items in 1945 and in 1965; calculate the percentage increase in the cost of these items:

	1945	1965
FOOD	$500	$650
RENT	$400	$800
CLOTHING	$100	$200

Solution

$$I = \frac{\Sigma p_n}{\Sigma p_o} \times 100$$

$$I = \frac{650 + 800 + 200}{500 + 400 + 100} \times 100$$

$$I = \frac{1{,}650}{1{,}000} \times 100$$

I = 165 or 165%

SIMPLE MEAN OF PRICE RELATIVES. The price relative is equal to the given period price, for an item, divided by its corresponding base period price. The MEAN of the price relative is therefore equal to the sum of all the price relatives divided by the number of items. Expressed symbolically:

$$I = \frac{\Sigma\left(\frac{p_n}{p_o}\right)}{K} \times 100$$

I = Index number

p_n = Given period price

p_o = Base period price

$\frac{p_n}{p_o}$ = Price relative (relative price)

$\Sigma\frac{p_n}{p_o}$ = Sum of all the individual price relatives for each item

K = Number of items

100 = Factor that converts the index number into a per cent.

1. SAMPLE "SIMPLE MEAN OF PRICE RELATIVES" PROBLEMS.

PROBLEM 1: Compute the mean of price relatives for the following data, using 1945 as the base period:

Item		1945 Average Price	1965 Average Price
Meat	(lb.)	$.25	$.75
Fish	(lb.)	.20	.60
Potatoes	(lb.)	.05	.15
Butter	(lb.)	.20	.80

Solution

$$I = \frac{\Sigma\left(\frac{p_n}{p_o}\right)}{K} \times 100$$

$$I = \frac{\frac{.75}{.25} + \frac{.60}{.20} + \frac{.15}{.05} + \frac{.80}{.20}}{4} \times 100$$

$$I = \frac{3+3+3+4}{4} \times 100$$

$$I = \frac{13}{4} \times 100$$

I = 325 or 325%

PROBLEM 2: Find the mean of the price relatives for the following data using 1945 as the base period:

Item	1945 Average Cost	1965 Average Cost
FOOD	$500	$650
RENT	$400	$800
CLOTHING	$100	$200

Solution

$$I = \frac{\Sigma\left(\frac{p_n}{p_o}\right)}{K} \times 100$$

$$I = \frac{\frac{650}{500} + \frac{800}{400} + \frac{200}{100}}{3} \times 100$$

$$I = \frac{1.3+2.0+1.0}{3} \times 100$$

$$I = \frac{5.3}{3} \times 100$$

I = 1.76 × 100

I = 176 or 176%

LASPEYRES WEIGHTED AGGREGATIVE PRICE INDEX. The two previous indexes gave equal treatment to all of the items being compared. In actual practice this is not true. For example, in the case of food, the average person consumes varying quantities of each food item. The price of meat is more important than the price of caviar. Therefore we should consider the quantities used as well as the price in computing a "realistic" index.

The Laspeyres weighted aggregative price index is equal to:

$$\frac{\Sigma \text{ (base period quantities of each item) times (correspon ding given period price)}}{\Sigma \text{ (base period quantities of each item) times (correspon ding base period price)}}$$

Expressed symbolically:

$$I = \frac{\Sigma(p_n q_o)}{\Sigma(p_o q_o)} \times 100$$

I = Index number

p_n = Given period price

q_o = Base period quantity

p_o = Base period price

100 = Factor which converts the index number into a percent.

1. SAMPLE LASPEYRES PRICE INDEX PROBLEMS:

PROBLEM 1: Compute the Laspeyres price index for the following data, using 1945 as the base period:

1945		
Item	Average Price	Average Quantity
Meat (lb.)	$ .25	100 lbs.
Fish (lb.)	.20	20 lbs.
Potatoes (lb.)	.05	100 lbs.
Butter (lb.)	.20	5 lbs.

1965		
Item	Average Price	Average Quantity
Meat (lb.)	$ .75	150 lbs.
Fish (lb.)	.60	30 lbs.
Potatoes (lb.)	.15	110 lbs.
Butter (lb.)	.80	5 lbs.

Solution

$$I = \frac{\Sigma(p_n q_o)}{\Sigma(p_o q_o)} \times 100$$

$$I = \frac{(.75\times100)+(.60\times20)+(.15\times100)+(.80\times5)}{(.25\times100)+(.20\times20)+(.05\times100)+(.20\times5)} \times 100$$

$$I = \frac{75+12+15+4}{25+4+5+1} \times 100$$

$$I = \frac{106}{35} \times 100$$

I = 302 or 303%s

PROBLEM 2: Compute the Laspeyres price index for the following data, using 1950 as the base period:

1950		
Item	Average Price	Average Quantity
Beef (lb.)	$ .60	50 lbs.
Pork (lb.)	.70	30 lbs.
Mutton (lb.)	. 50	20 lbs.

1965		
Item	Average Price	Average Quantity
Beef (lb.)	$ .50	70 lbs.
Pork (lb.)	.60	50 lbs.
Mutton (lb.)	.50	25 lbs.

Solution

$$I = \frac{\Sigma(p_n q_o)}{\Sigma(p_o q_o)} \times 100$$

$$I = \frac{(.50\times50)+(.60\times30)+(.50\times20)}{(.60\times50)+(.70\times30)+(.50\times20)} \times 100$$

$$I = \frac{25+18+10}{30+21+10} \times 100$$

$$I = \frac{53}{61} \times 100$$

I = 86.6 or 87%

The index number is less than 100. This means that there was a price decrease of 13%.

PAASCHE WEIGHTED AGGREGATED PRICE INDEX. The Paasche index is the same as the Laspeyres index except that in the Paasche index, given period quantities instead of base period quantities are used in the calculations. The formula for the Paasche index is therefore equal to:

$$I = \frac{\Sigma(p_n q_n)}{\Sigma(p_o q_n)} \times 100$$

I = Index number

p_n = Given period price

q_n = Given period quantity

p_o = Base period price

1. SAMPLE PAASCHE PRICE INDEX PROBLEMS.

PROBLEM 1: Compute the Paasche price index for the following data, using 1945 as the base period:

1945		
Item	Average Price	Average Quantity
Meat (lb.)	$ .25	100 lbs.
Fish (lb.)	.20	20 lbs.
Potatoes (lb.)	.05	100 lbs.
Butter (lb.)	.20	5 lbs.

1965		
Item	Average Price	Average Quantity
Meat (lb.)	$ .75	150 lbs.
Fish (lb.)	. 60	30 lbs.
Potatoes (lb.)	.15	110 lbs.
Butter (lb.)	. 80	5 lbs.

Solution

$$I = \frac{\Sigma(p_n q_n)}{\Sigma(p_o q_n)} \times 100$$

$$I = \frac{(.75\times150)+(.60\times30)+(.15\times110)+(.80\times5)}{(.25\times150)+(.20\times30)+(.05\times110)+(.20\times5)} \times 100$$

$$I = \frac{151}{50} \times 100$$

I = 302 or 302%

PROBLEM 2: Compute the Paasche price index for the following data, using 1950 as the base period:

1950		
Item	Average Price	Average Quantity
Beef (lb.)	$ .60	50 lbs.
Pork (lb.)	.70	30 lbs.
Mutton (lb.)	. 50	20 lbs.

1965		
Item	Average Price	Average Quantity
Beef (lb.)	$ .50	70 lbs.
Pork (lb.)	.60	50 lbs.
Mutton (lb.)	. 50	25 lbs.

Solution

$$I = \frac{\Sigma(p_n q_n)}{\Sigma(p_o q_n)} \times 100$$

$$I = \frac{(.50\times70)+(.60\times50)+(.50\times25)}{(.60\times70)+(.70\times50)+(.50\times25)} \times 100$$

$$I = \frac{35+30+12.50}{42+35+12.50} \times 100$$

$$I = \frac{77.50}{89.50} \times 100$$

I = 86.6 or 86.6%

FISHER'S IDEAL INDEX. Fisher's ideal index is equal to the square root of the product of Paasche's index multiplied by Laspeyres' index.

In other words:

$$\text{Fisher's index} = \sqrt{(\text{Paasche's index})(\text{Las peyres' index})}$$

$$\text{Fisher's index} = \sqrt{\frac{\Sigma(p_n q_n)}{\Sigma(p_o q_n)} \times \frac{\Sigma(p_n q_o)}{\Sigma(p_o q_o)}} \times 100$$

100 = Factor that converts index number into a percent.

1. SAMPLE FISHER' S IDEAL INDEX PROBLEMS:

PROBLEM 1: Compute Fisher's index for the following data, using 1945 as the base period:

1945		
Item	Average Price	Average Quantity
Meat (lb.)	$ .25	100 lbs.
Fish (lb.)	.20	20 lbs.
Potatoes (lb.)	.05	100 lbs.
Butter (lb.)	.20	5 lbs.

1965		
Item	Average Price	Average Quantity
Meat (lb.)	$ .75	150 lbs.
Fish (lb.)	. 60	30 lbs.
Potatoes (lb.)	.15	110 lbs.
Butter (lb.)	. 80	5 lbs.

Solution

$$I = \sqrt{\frac{\Sigma(p_n q_n)}{\Sigma(p_o q_n)} \times \frac{\Sigma(p_n q_o)}{\Sigma(p_o q_o)}} \times 100$$

$$I = \sqrt{(\text{Paasche's index})\ (\text{Laspeyres' index})}$$

From the previous problems, which we worked out in this topic, we know that for the above data, Paasche's index is equal to 303 and Laspeyres' index is equal to 303:

$$I = \sqrt{(302)(303)}$$

$$I = \sqrt{91{,}506} = 302.5$$

Note: We did not multiply by 100 because the index numbers had already been converted to percentages.

PROBLEM 2: Compute Fisher's index for the following data, using 1950 as the base period:

1950			
Item		Average Price	Average Quantity
Beef	(lb.)	$ .60	50 lbs.
Pork	(lb.)	.70	30 lbs.
Mutton	(lb.)	.50	20 lbs.

1965			
Item		Average Price	Average Quantity
Beef	(lb.)	$ .50	70 lbs.
Pork	(lb.)	.60	50 lbs.
Mutton	(lb.)	.50	25 lbs.

Solution

$$I = \sqrt{\frac{\Sigma(p_n q_n)}{\Sigma(p_o q_n)} \times \frac{\Sigma(p_n q_o)}{\Sigma(p_o q_o)}} \times 100$$

$$I = \sqrt{(\text{Paasche's index})\ (\text{Laspeyres' index})}$$

From the previous problems which we have worked out in this topic, we know that for the above data, Paasche's index is equal to 87 and Laspeyres' index is equal to 86.6

$$I = \sqrt{(87)(86.6)}$$

$$I = \sqrt{7534.2} = 86.8$$

END

INDEX

Absolute value
- minus signs and, 5
- notation for, 5

Addition
- law
  - combinations, 34-35
  - permutations, 30-35
- multiplication as form of, 10
- sign for, 10
- theorem (probability), 36

Aggregative price index, 84-85
- Laspeyres weighted, 85
- Paasche weighted, 86
- simple, 84-85

"And = multiplication" law, 33, 35
Arithmetic average, see Mean, the
Arrow notation, 6
Assumed averages, 15-23
- class intervals and, 17, 20, 23
- frequencies and, 15
- the mean and, 19

Average deviation, 5
- definition of, 5
- symbol for, 5

Bell shaped curves, 44
Between sample variance, 63, 65-69
Bimodals, 3
Binomial distribution, 40-41
- expansion, 40-41
- formula for, 40

Box method of calculating, 30
Chi square, 75
- pronunciation of, 75
- symbol for, 70

Chi Square Table, 75-76
Chi Square Test, 75
- formula, 75

Class boundaries, 17-18
Class intervals, 26
- assumed average and, 21
- frequency and
  - curves, 28
  - histograms, 26
  - polygons, 28
- graph, 26
- range of, 20

Class midpoint, 17, 20, 26-29
Combinations, 34-35
- addition law, 33-34, 36
- definition of, 34
- formula for, 33
- multiplication law, 33, 35-36

Confidence intervals
- large samples, 56
- small samples, 56

Confidence levels, 56
- Chi Square Tables and, 75
- T Tables and, 56-57

Correlation, coefficient
- equal to one, 39, 53-55, 70
- negative, 70
- positive, 7, 46, 70
- symbol for, 70

Curves,
- bell shaped, 44
- frequency
  - constructing, 7-8, 26-28
  - definition of, 2-3, 5, 7, 10, 36
- normal
  - table of areas under, 45

Data, see Grouped data, 19
Deviations
- average, 5
  - definition of, 5
  - symbol for, 5
- definition of, 5
- squaring, 7
- standard, 7
  - definition of, 7
  - of the difference between two means, 56-57
  - formula for, 7
  - frequency and, 10-11, 26-27
  - grouped data computation, 22
  - of the means, 56, 58, 64-68
  - normal curve, 44-46, 48-51
  - of the sample means, 56-58, 63, 66-67, 69
  - symbol for, 70

Distribution
- binomial, 40
  - expansion, 40
  - formula, 41
- frequency, 10-12

graphs, 10-12
histograms, 10-11
normal, 44-46, 56-58
areas of, 44-45, 50
curves, 44-45
of the sample means, 56-58
Poisson, 52-55
formula, 52
table of values, 52
students, 56
T, 56-62
"Either-or = addition" law, 33
Expansion, binomial, 40-41
F Table, 63-65, 67-69
F ratio, 63, 67-69
Factors
relationship between two, 80
symbol for, 70
Fisher, R. A., 63
Fisher's ideal index, 87
Frequencies
assumed average and, 15
Chi Square Test, 75
definition of, 75
distributions
graphs, 26, 28
histograms, 26
the mean and, 9-11, 13, 23, 44-45, 48, 50, 56
normal curve, 44-46, 48-51
polygons, 27
constructing, 27
definition of, 27
standard deviation and, 11
symbol for, 70
theoretical, 75
Gosset, W. S., 58
Graphs
of freqilency distributions, 26-29
curves, 28-29
histograms, 26
polygons, 27
frequency versus class interval, 26, 44
straight line, 28, 34, 79-83
Grouped data, 17
class intervals, 17-21, 23-24, 26-27
mean computation, 15
standard deviation computation, 17
Histograms, frequency
constructing, 7-8, 26-28
definition of, 2-3, 5, 7, 10, 36
normal curves, 44
Ideal index, Fisher's, 87
"If it is either one event or the other, then add" law, 36
"If it is, one event and the other event, then multiply" law, 37
Index numbers
Fisher's ideal index, 87
price indexes
Laspeyres weighted aggregative, 85
Paasche weighted aggregative, 86
simple aggregative, 84
simple mean of relatives, 85
symbol for, 70
Lambda (Greek letter), 52
Laspeyres weighted aggregative price index, 85
Laws
addition
combinations, 5, 34-35, 40-41
permutations, 5, 30-35
multiplication
combinations, 5, 34-35, 40-41
permutations, 5, 30-35
Least squares method, 80
Linear regression
definition of, 2-3, 5, 7, 10, 36
equation for a straight line, 79-83
least squares method, 15
Mean, the
assumed average and, 15
difference between more than two, 58
difference between two
normal deviate, 58
T Table, 56-60, 62
formula for, 1-2
frequency and,
grouped data computation, 19
large samples, 56
normal curve, 44-46, 48-51
numbers influence on, 2
of price relatives, 85
short cut method of calculating, 20
small samples, 56
standard deviation, 56
standard error of, 56, 58
uses for, 1
Median, the
definition of, 2
Minus signs
absolute value and,5
getting rid of, 7
Mode, the, 3
Multiplication
as a form of addition, 10
laws
combinations, 5, 34-35, 40-41
permutations, 5, 30-35
theorem (probability), 36
Normal curve
table of areas under, 45
Normal Deviate, 58
Normal distribution
areas of, 44-45, 50
curves, 44-51
of the sample: means, 56
Notations
absolute value, 5
arrows, 6
Numbers, index
Fisher's ideal index, 87
price indexes'
Laspeyres weighted aggregative, 85

Paasche weighted aggregative, 86
simple aggregative, 84
simple mean of relatives, 85
symbol for, 70
Paasche weighted aggregated price index, 86
Permutations, 30
addition law, 33-34, 36
box method of calculating, 30
definition of, 30
formulas for, 32-34
multiplication law, 33, 35-36
Poisson distribution
formula for, 52
table of values, 52
Polygons
definition of, 26
frequency
constructing, 27
definition of, 27
Populations, 56, 58
Price indexes
Fisher's ideal index, 87
Laspeyres weighted aggregate, 85
Paasche weighted aggregative, 86
simple aggregative, 84
simple mean of relatives, 85
Probability, 36
addition theorem, 36-37
binomial distribution computation, 40
Chi Square Table and, 75
definition of, 36
of an event not happening, 39
formula for, 36
multiplication theorem, 37-38
normal distribution computation, 44-51
Poisson distribution computation, 52
T Tables and, 58
Random sampling, 56
Range, the, 3
of the class, interval, 22-23
definition of, 3
Regression, linear
definition of, 2-3, 5, 7, 10, 36
equation for a straight line, 79-83
least square method, 80
Sample pair
analysis of variance, 5, 63, 65
confidence intervals, 56-57
difference between two means, 56, 58
Normal Deviate, 58
T Table, 56-60, 62
population, 56
Sampling, 56
confidence intervals, 56
small samples, 56
large samples, 56
definition of, 2-3, 5, 7, 10, 36
random, 56
standard error of the mean, 56
Sigma (Greek letter), 1
Squaring deviations, 7
Standard deviation, 5
definition of, 2-3, 5, 7, 10, 36
of the differences between two means, 58
formula for, 56
frequency and, 10-11, 26-27
grouped data computation, 22
of the means, 56, 58, 64-68
normal curve, 44-46, 48-51
of the sample means, 56-58, 63, 66-67, 69
symbol for, 70
Straight lines, 79
equation for, 79-83
graph, 79
least squares method, 80
Students' distribution, 56
Sum, sign for, 1
T distribution, 56
T ratio, 58-62
T Table, 56-60, 62
Theorems
addition, 7, 1, 10, 33-34, 36-37
multiplication, 10, 33, 35-38
Ungrouped data
mean computation, 20
standard deviation computation, 22
Value, absolute
minus sign and, 5
notation for, 5
Variance, the, 7
analysis of
F Table, 63-65, 67-69
formulas used in, 63
between sample, 58-59, 63, 65-69
definition of, 7
symbol for, 7
within sample, 63-69
Within sample variance, 63-69
X bar, 1

NOTES

CPSIA information can be obtained
at www.ICGtesting.com
Printed in the USA
FFOW01n0814130715
14959FF